Additive Manufactured Materials

This book discusses the thermo-mechanical and tribological behavior of various materials manufactured using additive manufacturing (AM). It examines a range of materials, including metals, polymers, ceramics, and composites, and provides insights into how processing parameters can be manipulated to achieve desired thermal, mechanical, and tribological characteristics. The book also addresses challenges related to the reproducibility and reliability of additive manufactured (AM) materials, offering solutions and strategies to overcome these hurdles in both research and industrial contexts. It does the following:

- Provides an understanding of the critical thermo-mechanical and tribological properties of AM materials
- Explains how thermal stress, heat treatment, and mechanical loading affect the performance of materials
- Offers a focused examination of friction, wear, and lubrication characteristics in AM materials
- Includes practical examples and case studies from industries such as the aerospace, automotive, and biomedical industries
- Discusses material performance and optimization strategies

This book is intended for graduate students and researchers in materials science, manufacturing, mechanical engineering, and industrial engineering.

Advances in Design, Materials and Manufacturing for Sustainability
Series Editors: Jitendra Kumar Katiyar and T V V L N Rao

Conventional and Powder Mixed Electro-Discharge Machining
Biomedical Applications
Edited by Ahmad Majdi Abdul-Rani, Masdi Muhammad, T V V L N Rao, Saeed Rubaiee, Anas Ahmed and Mohd Danish

Micro- and Biofluidics
Structure, Properties, Technology and Spotlight into the Future
Avinash Kumar and Jitendra Kumar Katiyar

Natural Fiber Composites for a Sustainable Environment
Edited by Jibin Philip, Jerin J Jose and Soney George

Material Selection, Product Design and Additive Manufacturing Processes for Operational Sustainability
Ram Krishna

Additive Manufactured Materials: Advances in Thermo-Mechanical and Tribological Peculiarity
Edited by Akant Kumar Singh, Naveen Mani Tripathi, Ankit Sharma, and Don McGlinchey

For more information about this series, please visit: www.routledge.com/Advances-in-Design-Materials-and-Manufacturing-for-Sustainability/book-series/CRCADMMS

Additive Manufactured Materials

Advances in Thermo-Mechanical and Tribological Peculiarity

Edited by
Akant Kumar Singh, Naveen Mani Tripathi,
Ankit Sharma, and Don McGlinchey

CRC Press
Taylor & Francis Group
Boca Raton London New York

CRC Press is an imprint of the
Taylor & Francis Group, an informa business

Designed cover image: Getty

First edition published in 2027
by CRC Press
2385 NW Executive Center Drive, Suite 320, Boca Raton FL 33431

and by CRC Press
4 Park Square, Milton Park, Abingdon, Oxon, OX14 4RN

CRC Press is an imprint of Taylor & Francis Group, LLC

ISBN: 9781041078531 (hbk)
ISBN: 9781041080558 (pbk)
ISBN: 9781003643531 (ebk)

DOI: 10.1201/9781003643531

Typeset in Times
by Newgen Publishing UK

Dedication

Dr. Akant Kumar Singh dedicates this book to Smt. Shiv Kumari Devi (mother) and Shree Balwant Singh (father).

Contents

Abbreviations

3D	three-dimensional
ABS	acrylonitrile butadiene styrene
ADM	atomic diffusion manufacturing
AFM	abrasive flow machining
AFM	atomic force microscopy
AI	artificial intelligence
ANFIS	adaptive Neuro-Fuzzy Inference System
ANN	artificial neural network
ANOVA	ANalysis Of VAriance
ASTM	American Society for Testing and Materials
ATC	anisotropic thermal conductivity
BIS	Bureau of Indian Standards
BJ	binder jetting
BN	boron nitride
CAS	computer-aided system
CAD	computer-aided design
CF	carbon fiber
CMC	ceramic matrix composites
CNC	computer numerical control
CrN	chromium nitride
CRPM	Centre for Rapid Prototyping and Manufacturing
CT	computed tomography
CVD	chemical vapor deposition
CVI	chemical vapor infiltration
DED	direct energy deposition
DfAM	Design for Additive Manufacturing
DIC	digital image correlation
DLC	diamond-like carbon
DLP	digital light processing
DMA	dynamic mechanical analysis
DMLS	direct metal laser sintering
DSC	differential scanning calorimetry
DTG	differential thermogravimetric analysis
EBM	electron beam melting
ECM	electrochemical machining
ECM	extracellular matrix
ECP	electrochemical polishing
EDM	electrical discharge machining
EDS	energy-dispersed X ray spectroscopy
EIS	electrochemical impedance spectroscopy
EP	electropolishing

EU	European Union
FAA	Federal Aviation Administration
FDA	US Food and Drug Administration
FDM	fused deposition modeling
FFF	fused filament fabrication
FML	fiber metal laminate
FOD	foreign object damage
GA-ANFIS	Genetic Algorithm-Adaptive Neuro-Fuzzy Inference System
GDP	gross domestic product
GE	General Electric
GHG	greenhouse gas
HA	hydroxyapatite
HCA	hydroxycarbonate
HCF	high-cycle fatigue
HEA	high entropy alloy
HIP	hot isostatic pressing
HVOF	high-velocity oxygen fuel
IEA	International Energy Agency
ILS	indirect laser sintering
IoT	Internet of Things
IPA	isopropyl alcohol
IPCC	Intergovernmental Panel on Climate Change
LAM	laser additive manufacturing
LCA	life cycle assessment
LCF	low-cycle fatigue
LST	laser-surface-textured
LOF	lack-of-fusion
LOM	laminated object manufacturing
LPBF	laser power bed fusion
LSR	laser surface remelting
MAAFM	magnetic-assisted AFM
MJF	multi-jet fusion
MJP	multi-jet printing
MMC	metal matrix composites
MRI	magnetic resonance imaging
MSE	mean squared error
MRAFF	magnetorheological AFM
MSLA	masked stereolithography
NCAM	National Center for Additive Manufacturing
OM	optical microscopy
PA	polyamide
PBF	powder bed fusion
PC	polycarbonate

PCL	poly-caprolactone
PEEK	polyether ether ketone
PEI	polyetherimide – ULTEM
PEKK	polyether ketone ketone
PEM	proton exchange membrane
PET	polyethylene terephthalate
PETG	polyethylene terephthalate glycol
PH	precipitation-hardening
PID	proportional integral derivative
PIP	polymer infiltration and pyrolysis
PLA	polylactic acid
PMC	polymer matrix composites
PP	polypropylene
PPSU	polyphenylsulfone
PVA	polyvinyl acetate
PVD	physical vapor deposition
RSM	response surface methodology
SAED	selected area electron diffraction
SAG	shape-adaptive grinding
SEAM	surface engineering and advanced materials
SEBS	styrene ethylene butadiene styrene
SEM	scanning electron microscopy
SLCA	social life cycle assessment
SLA	stereolithography
SLM	selective laser melting
SLS	selective laser sintering
SME	small and medium enterprise
SPD	severe plastic deformation
STEM	science, technology, engineering and mathematics
STM	scanning tunneling microscopy
TCP	tricalcium phosphate
TDS	thermal decomposition of salts
TEM	transmission electron microscopy
TGA	thermogravimetric analysis
TiN	dense titanium nitride
TPE	thermoplastic elastomers
TPU	thermoplastic polyurethane
UHMWPE	ultra-high molecular weight polyethylene
UNEP	United Nations Environment Program
UTM	universal testing machine
UV	ultraviolet
VOC	volatile organic compound
XCT	X-ray computed tomography

UTS	ultimate tensile strength
WEF	World Economic Forum
XRD	X-ray diffraction
YSZ	Yttria-stabilized zirconia
VC	vanadium carbide
VPP	vat photopolymerization

Editors' Biographies

Akant Kumar Singh graduated in Mechanical Engineering from Dr. A.P.J. Abdul Kalam Technical University Lucknow, Uttar Pradesh, India. He has finished his master's and Ph.D. degrees in Mechanical Engineering from the National Institute of Technology Hamirpur, Himachal Pradesh, India. His Ph.D. research is on the topic of "tribological aspects of the fiber augmented functionally graded material-based polymer spur gears". He has evolved a novel mathematical equation to calculate the wear rate of polymer gears. He has 14 years of teaching and research experience. He has 37 research publications in international journals and 12 publications in international conferences. He has published seven book chapters with reputed international publishers. He has also published one Indian patent. He is a reviewer of various international journals. His research interest lies in the area of functionally graded materials (FGMs), polymer gears, and materials tribology. He has guided one master's thesis. Five students are pursuing their master's thesis work under his guidance. Currently, he is working as an Assistant Professor in the Department of Mechanical Engineering at Chandigarh University Mohali, Punjab, India.

Naveen Mani Tripathi is currently working as an Assistant Professor in the Department of Mechanical Engineering at Rajiv Gandhi Institute of Petroleum Technology, Jais, Amethi, Ministry of Petroleum and Natural Gas, Government of India. He completed his Ph.D. from the Mechanical Engineering Department of Ben Gurion University of the Negev, Beer-Sheva, Israel. After his Ph.D., he worked as a particle scientist in the material characterization instrument manufacturing company, GranuTools sprl, Belgium. His postdoctoral research is from the Chair of Process System Engineering of the Technical University of Munich, Germany. He has been awarded many fellowships, such as the Graduate Aptitude Test in Engineering (GATE), the Graduate Record Examination (GRE), the Kriteman Fellowship, the Technical University of Munich's University Foundation Fellowship, and the Alexander von Humboldt Fellowship. His area of research is material characterization, additive and smart manufacturing, and multiphase flow. He has four granted and one published patent. He has published more than 50 journals/conference publications and is the editor of three books from international publishers. He is voluntarily working as an editorial board member and review editor for more than 20 journals. He is Associate Editor of the Scopus-indexed journal *Materials Process Engineering*, *Frontiers in Chemical Engineering*. He has also served as the Editor of a special issue of the journal *Recent Patents on Mechanical Engineering*, *Betham Science*.

Ankit Sharma holds the position of Associate Director (Research) of the Research and Innovation Network (CURIN) Department, Chitkara University, Punjab, India. He received his doctorate in Mechanical Engineering from the Thapar Institute of Engineering and Technology, India. He has more than 10 years of experience in academics, research, consulting, training, and industry. Dr. Sharma also organized an international conference – the International Conference on Emerging Materials,

Smart Manufacturing and Computational Intelligence (ICEMSMCI) 2023 and served as the General Chair and Convenor. He has authored numerous national and international publications in Science Citation Index (SCI), Scopus, and Web of Science-indexed journals. He has filed/published more than 30 international patents. He is the editor of several CRC Press, Taylor & Francis Group publications related to advanced manufacturing processes, and several are in the pipeline.

Don McGlinchey is the academic leader of the Digital Transformation and Intelligent Maintenance Research Group and the former Head of the Centre for Industrial Bulk Solids Handling at Glasgow Caledonian University. He is an academic of international standing within the particulate solids handling community with over 30 years' experience. He has undertaken contract research and consultancy projects for both multinational companies, and small to medium enterprises (SMEs), and delivered short courses in Europe, the USA and the UK. He is a chartered physicist, with a Ph.D. in the particle technology area. He is the editor of three books and has authored over 50 research articles. His current research interests include multiphase flow pipeline transport measurement and modelling, and pneumatic conveying system modeling.

Contributors

Anant Prakash Agrawal
Department of Mechanical Engineering
Noida Institute of Engineering and Technology
Greater Noida, Uttar Pradesh, India

Mohamed Sukri Mat Ali
Universiti Teknologi Malaysia (UTM)
Kuala Lumpur, Malaysia

Shahazad Ali
Department of Mechanical Engineering
Noida Institute of Engineering and Technology
Greater Noida, Uttar Pradesh, India

Mohd Imran Ansari
Department of Robotics and Automation Engineering
Bharati Vidyapeeth (DU) College of Engineering, Pune
Maharashtra, India

Madhusudan Baghel
Laboratory for High Efficiency and Clean Mechanical Manufacture of Ministry of Education
School of Mechanical Engineering
Shandong University
Jinan, China

Borad M. Barkachary
Department of Mechanical Engineering
Jorhat Institute of Science and Technology
Jorhat, Assam, India

Sachin S. Chavan
Department of Mechanical Engineering
Bharati Vidyapeeth (DU) College of Engineering, Pune
Maharashtra, India

Jasgurpreet Singh Chohan
Department of Mechanical Engineering
Rayat Bhara University, Punjab, Mohali
Punjab, India

Prachi A Dixit
Department of Mechanical Engineering
Bharati Vidyapeeth (DU) College of Engineering, Pune
Maharashtra, India

Navdeep Sharma Dugala
Department of Automobile Engineering
Chandigarh University, Mohali
Punjab, India

Juan Francisco Resendiz García
Automotive System Engineer, Sulzer Chemtech
Autopista México – Querétaro 57d, Industrial Cuamatla
54730, Cuautitlán Izcalli, Méx., Mexico

Arun Gupta
Department of Mechanical Engineering
Teerthanker Mahaveer University, Moradabad
Uttar Pradesh, India

Kashif Irshad
Interdisciplinary Research Centre for Sustainable Energy Systems (IRC-SES)
Research Institute, King Fahd University of Petroleum and Minerals
Dhahran, 31261, Saudi Arabia

Amrinder Singh Johal
Department of Automobile Engineering
Chandigarh University, Mohali
Punjab, India

Deepak Kapila
Department of Automobile Engineering
Chandigarh University, Mohali
Punjab, India

Akhilesh Kumar
Department of Civil Engineering
Chitkara University Institute of Engineering and Technology
Chitkara University, Punjab

Ashish Kumar
Ministry of Education Key Lab for Liquid-Solid Structure Evolution and Materials Processing
Shandong University
Jinan, China

Dinesh Kumar
Department of Mechanical Engineering
Teerthanker Mahaveer University, Moradabad
Uttar Pradesh, India

Shailendra Kumar
Satyug Darshan Institute of Engineering and Technology
Faridanad, Haryana

Surender Kumar
Department of Mechanical Engineering
Chandigarh University, Mohali
Punjab, India

Ribhu Kundu
Department of Automobile Engineering
Chandigarh University, Mohali
Punjab, India

Ritesh G. Mahajan
Department of Robotics and Automation Engineering
Bharati Vidyapeeth (DU) College of Engineering, Pune
Maharashtra, India

Rahul Nadda
Department of Chemical Engineering
Loughborough University, Loughborough
LE11 3TU, United Kingdom

Mahesh A Naik
R&T Scientist
Functional Polymer Unit
LIST, Luxembourg

Mandeep Pathak
Department of Mechanical Engineering
Jorhat Institute of Science and Technology
Jorhat, Assam, India

Anant Krishan Pun
Department of Mechanical Engineering
Chandigarh University, Mohali
Punjab, India

Gagandeep Roop Rai
Department of Mechanical Engineering
Chandigarh University, Mohali
Punjab, India

Reddyvari Loka Vigneswar Reddy
Department of Automobile Engineering
Chandigarh University, Mohali
Punjab, India

Sulaim Rizvi
Department of Mechanical Engineering
Noida Institute of Engineering and Technology
Greater Noida, Uttar Pradesh, India

Ahmed Fakhrudin Safdari
Department of Mechanical Engineering
University of Central Florida
Florida, USA

Satbir Singh Sehgal
Department of Mechanical Engineering
Chandigarh University, Mohali
Punjab, India

Utpal Sen
Department of Mechanical Engineering
Jorhat Institute of Science and Technology
Jorhat, Assam, India

Vinay Shah
Department of Mechanical Engineering
Chandigarh University, Mohali
Punjab, India

Mufaddal Huzefa Shakir
Department of Materials and Processing
University of Waikato
Hamilton, New Zealand

Siddhartha Sharma
Department of Mechanical Engineering
National Institute of Technology Hamirpur
Himachal Pradesh, India

Areeb Shazli
Department of Automobile Engineering
Chandigarh University, Mohali
Punjab, India

Akant Kumar Singh
Department of Mechanical Engineering
Chandigarh University, Mohali
Punjab, India

Balwant Singh
Department of Mechanical Engineering
Chandigarh University, Mohali
Punjab, India

Bikramjeet Singh
Department of Mechanical Engineering
Chandigarh University, Mohali
Punjab, India

Jaspreet Singh
Department of Automobile Engineering
Chandigarh University, Mohali
Punjab, India

Manpreet Singh
Department of Mechanical Engineering
University of Maryland, Baltimore County
USA

Prashant Kumar Singh
Department of Mechanical Engineering
Chandigarh University, Mohali
Punjab, India

Virendra Pratap Singh
Ministry of Education Key Lab for Liquid-Solid Structure Evolution and Materials Processing
Shandong University
Jinan, China

Naveen Mani Tripathi
Department of Mechanical Engineering
Rajiv Gandhi Institute of Petroleum Technology, Jais
Amethi, Uttar Pradesh, India

Manvendra Yadav
AIC-EMPI Incubation Foundation
New Delhi, India

Sanjay Yadav
Department of Mechanical Engineering
ITS Engineering College, Greater Noida
Uttar Pradesh, India

Preface

Additive manufactured materials and the advances in thermo-mechanical and tribological peculiarity have acquired considerable importance in recent years. This book will serve as an important resource for researchers interested in this and related fields of engineering. This extensive text provides essential and systematic insights into topical tribology issues as an aid to a better understanding of the unique thermo-mechanical and tribological properties of additive manufactured (AM) parts.

This book provides a high-level overview of cutting-edge developments and in-depth knowledge of materials selection, processing technology, and characterizations for performance improvements required to deliver next-generation smart material systems. Readers will learn about the myriads of materials, from metals and polymers to ceramics and composites, how this presents a challenge for selecting one that can withstand complex thermal processes involving simultaneously high temperatures (over 500°C) generally suitable with metal processing, and how they can be optimized for superior performance in industries such as aerospace, the automotive industry, and biomedical engineering.

The book aims to be at the forefront of additive manufacturing research and application. This is a must-read for anyone seeking to understand the future of materials science and engineering.

Acknowledgment

We express our sincere thanks to all contributors, all those friends and colleagues who have activated our norms to take up a study like this.

The editors greatly remember the contribution and support provided by Shri Satnam Singh Sandhu, the Honorable Chancellor, Chandigarh University, Punjab; Prof. Harish Hirani, the Honorable Director, Rajiv Gandhi Institute of Petroleum Technology; Dr. Sandhir Sharma, the Honorable Vice-Chancellor, Chitkara University, Punjab; and Prof. Steve Decent, the Honorable Vice-Chancellor, Glasgow Caledonian University, United Kingdom, for providing a vibrant research and academic ecosystem.

We are highly indebted to our parents, spouses, children and parents-in-law, whose support and advice have paved the way for our study and investigations. They have created a congenial atmosphere at home for our studies and writing. They are silent observers of our works.

Dr. Akant Kumar Singh
Dr. Naveen Mani Tripathi
Dr. Ankit Sharma
Prof. (Dr.) Don McGlinchey

1 Introduction to Materials Used for Additive Manufacturing

Surender Kumar, Akant Kumar Singh, Naveen Mani Tripathi, Akhilesh Kumar, and Juan Francisco Resendiz García

1.1 INTRODUCTION

Advanced manufacturing (AM) techniques are important to fulfil the required demands of society. The manufacturing process should also be rapid to decrease the lead time. Additive manufacturing is one of the advanced manufacturing techniques used to manufacture components in less time and with minimum wastage of raw materials. A lot of research has been going on in the field of AM to manufacture a wide range of products with a wide range of materials. The selection of material is an important criterium for adopting the correct AM techniques (Rajendran et al., 2023; Sadaf et al., 2024). Broadly, material used in the rapid manufacturing (RM) process can be categorized into three types:

- Polymers
- Metal
- Ceramics

1.1.1 Importance in Modern Manufacturing

The need for advanced manufacturing techniques is required for manufacturing the products or components in less time and with high accuracy. The AM techniques have many applications in different fields of industry. In AM, the components are directly manufactured from a CAD drawing by adopting the right type of AM process. The traditional subtracting manufacturing process is a time-consuming process and needs a number of operations to produce the final components. An AM process enables one to manufacture the component with a single process. As the requirements of components are increasing as time passes, so we require processes that can develop the required components in less time and at high accuracy. As per the American Society for Testing Material (ASTM)-F42, AM techniques are classified into seven categories according to their methods of process (Kawalkar et al., 2022). Additive manufacturing is one of the most cost-effective techniques. At the same time, adopting it for mass production is still a challenge. AM processes are most effective for components

DOI: 10.1201/9781003643531-1

that require frequent design modifications as per requirements (Ahuja et al., 2015; Tofail et al., 2018).

1.1.2 Importance of Material Selection

The success of AM techniques depends upon the selection of appropriate materials. This is the most important factor in the AM process. The incorrect selection of material can lead to defects and failure, while the right selection of material ensures efficient and reliable components. The following points are important for material selection (Agrawal, 2021, Alghamdy et.al., 2019):

- Material performance: The durability, strength, and functionality of the developed components depends directly upon material selection.
- Process compatibility: Every AM technique supports only certain materials; the wrong selection of material may lead to failure or defects in the developed components.
- Accuracy and finish: The properties of the selected material influence the surface quality or dimensional accuracy of the developed components.
- Application suitability: The selection of material for AM also depends on the application of the component for which it is developed. Different fields (aerospace, automobile, aeronautical fields) demand specific properties like lightweight, high-strength, high-temperature registered materials.
- Sustainability: Eco-friendly material supports green manufacturing techniques in order to minimize environmental impact and still meet the performance requirements.

1.1.3 Scope of This Chapter

The fundamentals of AM with regard to materials like polymers, metals, and ceramics are discussed in this chapter. The present book chapter aims to present the basic ideas of AM and material selection. The processes and material relationships are discussed in this chapter. The physical and chemical nature of polymers, ceramics and metals, and how they play a role in the selection of AM processes, are discussed thoroughly.

The requirements of post-processing techniques related to performance requirements and industrial applications are included in the present chapter. The future scope in the field of AM techniques and their materials is also presented.

1.2 BASIC PROCESS OF ADDITIVE MANUFACTURING

Polymer is the most suitable material for AM polymer materials. Additive manufacturing can be performed easily in less time, and versatile products can be manufactured using polymer material. Polymer material requires a lower operating temperature, so it does not require machinery that needs to be operated at a high

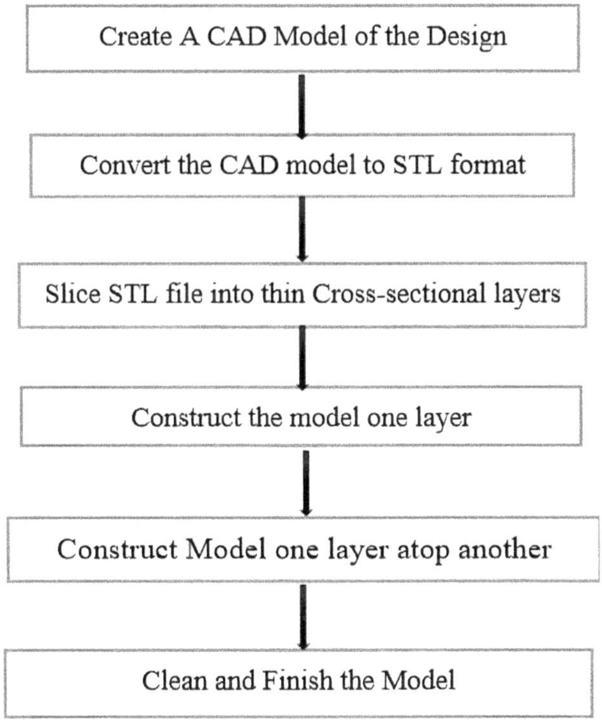

FIGURE 1.1 Additive manufacturing process chart.

temperature. Polymer material was the first material to be used for AM processes in 1984.

The basic process of AM for polymer material is the same as for other AM processes. CAD software is used to create a three-dimensional (3D) model of the required component. The developed CAD model is then converted into a standard tessellation language (STL) file by a tessellation process. Slicing software then converts the CAD model into layers. The selected AM process then prints the component as per the CAD model (Kumar et al., 2021; Negi et al., 2013). The post-processing of printed components performed to achieve the required dimension of the final components is given in Figure 1.1.

1.3 MATERIAL OF ADDITIVE MANUFACTURING

The additive manufacturing process started in the mid-1980s. A lot of research has been done on materials and techniques for the AM process. In this chapter, materials of additive manufacturing are discussed (Huang et al., 2013; Jiang et al., 2018; Wong & Hernandez, 2012). There are specific AM techniques that can only work with specific materials. They are also discussed in this chapter.

1.3.1 Polymers (Thermoplastics, High-performance Polymers, Photo-resins)

Polymer is the foundation of AM because the first AM material was a polymer. Polymer material is used for functional components and for prototyping. The prototyping affects the product's development life cycle when launching and adding new features to the product as per the demand of the customer (Chatterjee et al., 2023). Polymer material can be divided into three categories.

1.3.1.1 Thermoplastics

Thermoplastic is a type of material that becomes soft by heating, and hard when it cools down. Such properties make it suitable for AM and for recycling (Tan et al., 2020). Thermoplastic materials are mainly used for selective laser sintering (SLS), fused filament fabrication (FFF) and multi-jet fusion (MJF).

- Polylactic Acid (PLA): This material is used mostly in FFF and fused deposition modeling (FDM) AM. Material is developed from corn starch, sugarcane and tapioca roots. PLA has developed as a biodegradable thermoplastic with application in prototyping, education, and consumer products. PLA is easy to process, and qualities like low warping and good printability make it highly suitable for AM industries. Its brittleness and poor thermal resistance are the limitations that restrict its use for high-temperature applications. PLA is increasingly being explored for biomedical skeletons (Singh et al., 2020).
- Acrylonitrile Butadiene Styrene (ABS): This is one of the most widely used AM material for FDM/FFF. ABS is an excellent material for AM due to its combination of toughness, strength, and operability. ABS has high thermal stability compared to PLA. ABS has applications in housing, automobiles, consumer electronics casings, and functional prototyping. Post-processing can also be performed easily on ABS. The recycling of ABS is also possible, which makes it a sustainable material for AM (Vidakis et al., 2020).
- Polyethylene Terephthalate Glycol (PETG): This material is used for FDM/FFF AM processes. PETG has excellent chemical resistance and toughness. This material can be used for medical devices, watches, and for packaging. PETG provides good printability compared to ABS, and it is more durable than PLA (Hassan et al., 2020).
- Polyamides (nylon): This material is used for SLS and MJF processes. It has good strength, wear-resistance, and chemical-resistance. This material also provides flexibility in the development of components. Gears, functional automotive parts, and snap-fit components are printed using polyamides.
- Polycarbonate (PC): In additive manufacturing, PC is mostly used for FDM/FFF processes, and to a much less extent, in SLS. PC is used for AM because it has heat stability, impact resistance, and excellent toughness. PC is broadly applied in products that require the thermal resistance of functional parts, high strength, and durability.

1.3.1.2 High-performance Polymer

High-performance polymers are advanced engineering thermoplastic materials developed for demanding applications in the defense, energy, medical, automobile, and aerospace sectors. Compared to other polymer materials, they have excellent heat resistance, chemical stability and biocompatibility. They require AM techniques that make use of high operating temperatures (Boydston et al., 2020; Kafi et al., 2020).

- Polyether Ether Ketone (PEEK): This material is used for FFF/FDM and SLS techniques at high temperatures. It has application in the medical, aerospace, and energy industries. PEEK has excellent strength, thermal stability, chemical resistance, and wear resistance. Its properties are comparable with metal, so it is in high demand to replace metals.
- Polyether Ketone Ketone (PEKK): This is suitable for FFF/FDM and SLS AM processes. PEKK is similar to PEEK, but it has a slower crystallization rate and a flexible molecular structure, which makes it easier to process by AM. PEKK has a higher tensile strength and warping tendency compared to PEEK. PEKK has application in the fields of aerospace, defense, and medicine. This material is easy to print because it has a slower crystallization rate.
- Polyetherimide (PEI): This is suitable for FFF/FDM and SLS AM processes. PEI, commercially known as ULTEM, is a thermoplastic material used extensively in industrial AM. PEI is cost-effective compared to PEEK and PEKK. At the same time, it has good thermal, mechanical, and frame retardant properties. PEI's balance of heat resistance, inherent flame retardancy, and strength makes it highly suitable for automotive, aerospace, and electronic applications.

1.3.1.3 Photo-resins

Photo-resins are light-sensitive liquids. They experience photo-polymerization when exposed to a light source (laser or UV). The chemical reaction occurs in the photo-resin when exposed to light. In AM, photo-resins serve as the main material for vat photopolymerization (VPP) processes, i.e. stereolithography (SLA), poly-jet/multi-jet printing (MJP), LCD masked stereolithography (MSLA) and digital light processing (DLP). Once solidified with photopolymerization, these materials cannot be remelted or reshaped. They materials have application in the fields of dental medical devices.

1.3.2 Metals and Alloys for Additive Manufacturing

Metals and alloys for AM are modern manufacturing methods. These methods directly produce components rather than prototypes, as in the case of polymers. The AM applications of metals are in medical hardware, aerospace materials, tool and die moulds, automotive and remanufacturing parts, etc. The AM of metallic parts with complex geometries has attracted tremendous attention (DebRoy et al., 2018).

1.3.2.1 Stainless Steel

Stainless steel is a metallic material used mostly for AM processes like powder bed fusion (PBF), selective laser melting (SLM), electron beam melting (EBM), direct energy deposition (DED) and binder jetting. Stainless steel has important properties like corrosion resistance, high mechanical strength, and design flexibility, making it suitable for aerospace and medical tool applications (Sing et al., 2016).

- Austenitic stainless steels (Common Grade: 316L): These are used for PBF and DED RM techniques. These are steels with high corrosion resistance and excellent mechanical properties. The material is biocompatible, hence it can be used for surgical tools. 316L stainless steel has application in chemical industry and food processing equipment, and surgical instruments.
- Precipitation-hardening (PH) stainless steels (Common Grade 17-4PH): This material offers high strength (1000 Mpa) and corrosion resistance. Precipitation hardening can be done on stainless steel to achieve the required strength and ductility balance. It is used in nuclear parts, marine components, and aerospace tooling.

1.3.2.2 Titanium and Its Alloys

Titanium and its alloys are among the most important materials for AM, especially in the aerospace, defense, and biomedical industries. SLS, EBM and DED are the AM techniques that can be performed with titanium alloys. Their corrosion resistance, high strength-to-weight ratio, and biocompatibility make them ideal for producing complex, lightweight, and patient-specific components. The AM of a titanium alloy is easy compared to subtractive manufacturing due its high strength and poor machinability (Tshephe et al., 2022).

- Ti-6Al-4V (Grade 5): This is the most commonly used titanium alloy for AM processes. This alloy is employed in EBM and laser power bed fusion (LPBF) processes. The Ti-6Al-4V alloy is used in orthopedic implants, aerospace brackets, and dental prosthetics. Titanium's biocompatibility and low modulus (~110 GPa) make it the ultimate alloy for bone implants. Research on titanium alloys is ongoing to find the best material for AM and for specific applications (Qian et al., 2016).

1.3.2.3 Aluminum Alloys

Aluminum (Al) and its alloys are used for PBF and DED AM techniques. Aluminum and its alloys have many important properties, like high specific strength, excellent corrosion resistance and low density, which make it suitable for many industrial applications. These materials are gaining prominence for use in the automotive, aerospace, and defense industries, and in consumer electronics.

- AlSi10Mg: AlSi10Mg is a hypoeutectic aluminum silicon alloy with 0.3wt.% Mg and 10%Si. AlSi10Mg is a castable alloy especially designed for AM with a high corrosion resistance and high fatigue strength. This alloy is used for LPBF AM techniques due to its possibility and availability. Automotive, housing, and lightweight prototypes can be printed using AlSi10Mg.

- Scalmalloy (Al-Mg-Sc-Zr): Scalmalloy is a high-performance Al-Mg-Zr alloy especially developed for AM. It combines the light weight of aluminum with strength comparable to titanium alloy. Scalmalloy is the strongest aluminum alloy available for AM, and is particularly valuable in aerospace, defense, and motorsport applications.

1.3.2.4 Nickel-based Superalloys

In AM, PBF and DED processes, in particular, can be performed using a nickel-based superalloy. These alloys have a high strength, and are corrosion resistant and creep resistant at elevated temperatures. They are the pillar of nuclear systems, power generation turbines, and aerospace engines. Complex geometry using these alloys can be developed by AM processes that are not possible using subtractive manufacturing processes.

- Inconel 718 and Inconel 625: These materials can withstand high temperatures (605°C). They are used in high-temperature applications like gas turbines, jet engines, and rocket combustion chambers. These alloys are also suitable for chemical plants and the marine environment.

Metal and alloys are found to be suitable for structural, biomedical, and thermal applications where polymers cannot be use for such applications due to their thermal strength and chemical properties.

1.3.3 Ceramics for Additive Manufacturing

Ceramic materials are among the most challenging, but very valuable materials in AM. They are recognized for their wear resistance, hardness, corrosion resistance and high temperature stability. Ceramics have many applications in the biomedical, aerospace, energy, and electronics industries. Ceramics are a highly brittle material, and are difficult to process due to their high melting point (Deckers et al., 2014; Zocca et al., 2015).

1.3.3.1 Technical Ceramics

Ceramics are used in the defense, aerospace, and energy sectors due to their thermal and mechanical properties.

- Aluminum oxides (Al_2O_3): Aluminum oxide, also known as alumina, is the most important ceramic oxide. In AM, many researchers have worked on Al_2O_3 ceramics due to their good printability in both slurry and powder form. Alumina can be used in liquid- and powder-based AM techniques, and have application in electrical insulators, biomedical implants, and cutting tools due to their wear resistance, electrical insulation and biocompatibility.

- Zirconium dioxide (ZrO_2): Zirconium dioxide is also known as zirconia. In AM, it is used in SLA, SLS, binder jetting (BJ) and extrusion techniques. In AM, it is used due to its exceptional chemical, mechanical, and thermal properties. It is widely used in high-strength, wear-resistant and biocompatible applications.
- Silicon carbide (SiC): Silicon carbide is also a ceramic material that is used in AM techniques as a raw material. It is compatible with BJ, material extrusion and SLS techniques. Still, the processing of SiC is challenging due to its high melting point and brittleness. It is used for heat exchanger and turbine components, and mechanical seals.

1.3.3.2 Bio-ceramics

Bio-ceramics are useful in dentistry and orthopedics because they have good biocompatibility. They have a wide range of applications in medical fields because they are biogenerated.

- Hydroxyapatite (HA, $Ca_{10}(PO_4)_6(OH)_2$): This is a material that looks a lot like the mineral phase of human teeth and the bone of the human body. It is the most biocompatible material for both the human body and teeth. This material has excellent bioactivity, which forms a direct bond between the bone and tissue of the human body. The material is brittle, so it cannot be used for high load-bearing applications. With AM process like SLS, SLA and BJ, it is used to fabricate bone scaffolds, dental implants, and metallic implants (García-Hernández et al., 2025).
- Tricalcium phosphate (TCP): This material is used for biomedical applications due to its biocompatibility. It is widely used in the case of bone regenerations due to its bioresorbability and biodegradability under physiological conditions. TCP easily dissolves in human body fluid and is replaced by new bone through natural remodeling. This material is an ideal material for bone scaffolds. The mechanical strength of TCP is lower compared to hydroxyapatite (HA). AM processes such as extrusion-based printing and SLS enable the development of highly porous TCP scaffolds with the required geometry and mechanical strength. HA and TCP are used in combination to achieve the required strength and stability (Tarafder & Bose, 2013).
- Bioactive glasses (e.g. 45S5 Bioglass): Bioglass is another class of bioactive material used in tissue and bone engineering. A composition of bioglass is CaO, Na_2O, SiO_2 and P_2O_5. These materials have the ability to develop hydroxycarbonate (HCA) on the surface when exposed to physiological fluids. Bioactive glass not only speeds up osseointegration, but also simulates osteogenesis and angiogenesis. The brittle nature of bioactive glass limits its use in high load-bearing applications. In AM, bioglasses are developed by using FDM with polymer binders and inkjet printing techniques. They are mostly used in bone defect treatment and coating on metallic implants (Santos et al., 2024).

1.4 COMPARATIVE ANALYSIS, CHALLENGES AND FUTURE TRENDS

The first AM with polymer-based material was done by Charles W. Hull in 1986 using the SLA technique. The research into advanced materials for AM is ongoing to find new advanced materials for required applications. The use of metallic material for AM techniques started in the mid-1990s after the advent of SLM and EBM techniques. Advanced metallic materials have been developing as per applications. Ceramic-based AM materials were used in 2000 with the adaptation of binder jetting, robocasting and SLA with HA, TCP, ZrO_2, and Al_2O_3. Ceramic-based materials exhibit high hardness, wear resistance and also have biomedical applications (Ngo et al., 2018).

Polymer materials offer greater design flexibility, a low process temperature and biocompatibility, but possess lower thermal and mechanical strength. Metallic materials provide high thermal conductivity, mechanical strength and corrosion resistance, supporting applications in the automotive and aerospace industries, and in medical implants. Ceramic materials provide supreme hardness, thermal stability, and bioactivity, but are brittle and challenging to process. The processing of AM material is still a challenging task to achieve the required properties of the developed components. Polymers have thermal instability, low mechanical strength and anisotropic mechanical properties due to their layer-by-layer fabrication process. The metals encounter issues like porosity and residual stress formation, and need post-processing like hot isostatic pressing (HIP) or heat treatment to achieve the required density and microstructure. The processing temperature of metallic materials is very high due to high-temperature manufacturing with metallic material requiring special AM methods and machines. Ceramics pose extreme manufacturing challenges due to their high melting point, shrinkage sintering, and inherent brittleness. Due to their brittleness, ceramics are prone to crack and form dimensional inaccuracies. In all three materials (polymers, metals and ceramics), process standardization, repeatability, and quality assurance continue to be major issues to adopt them for industrialization (Ford & Despeisse, 2016).

The future materials for AM need to overcome these limitations in order to develop novel materials. The advanced material formulations' smart processing techniques and hybrid systems can be used to overcome these limitations. In polymers, the development of biodegradable materials, as well as the use of recyclable and fiber-reinforced composites, can be used. The addition of carbon nanotubes in polymers can provide good mechanical strength for materials. Metal based AM is evolving towards multi-material printing, and *in-situ* monitoring systems in order to develop components with a good strength and surface finish. Artificial intelligence (AI) can be used in metallic AM techniques to achieve specific control over the process. Ceramic AM research emphasizes using photo-curable ceramic slurry, ceramic polymer composites, and low temperature sintering additives to achieve high strength and low brittleness, and enhance structural reliability (Liu et al., 2022). Table 1.1 depicts the advent of material for AM, along with their processes and materials.

TABLE 1.1
Overview of Additive Manufactured Materials

Sr. No	Material type	Period of introduction/ development	Key AM processes	Common materials	Typical applications
1.	Polymers	1986 to present (SLA introduced by Charles W. Hull, 1986; FDM by Scott Crump, 1989)	SLA, FDM, SLS, MJF	ABS, PLA, PETG, nylon (PA12), PEEK, PEI (ULTEM), photopolymers	Prototyping, biomedical scaffolds, consumer products, educational models
2.	Metals	1991 to present (SLS patented by Deckard, 1991; SLM/ EBM matured by 2005)	SLM, EBM, DED, binder jetting	Ti-6Al-4V, 316L SS, Inconel 718, AlSi10Mg, Co-Cr alloys, high-entropy alloys	Aerospace parts, automotive components, implants, tooling
3.	Ceramics	2000 to present (ceramic AM through robocasting and SLA around 2000–2010)	Binder jetting, SLA, robocasting, inkjet printing, laminated object manufacturing	HA, TCP, zirconia (ZrO_2), alumina (Al_2O_3), bioactive glasses (45S5)	Biomedical implants, dental prosthetics, electronic substrates, thermal barrier coatings

1.5 CONCLUSION

The materials used for AM play an important role in the strength, properties, and longevity of developed AM materials. The evolution of AM has been strongly influenced by the continuous development of new or modified materials for different AM techniques. The initial polymers have been used for AM techniques. Due to their limited use and strength, metallic material is also used for AM. Ceramic materials are the most recent category of materials that has been used for AM. Each category of material has its own unique chemical, mechanical, and physical properties, enabling lightweight structures, complex geometries and high-performance components for multiple industries.

Polymers are the most cost-effective material, which is the easiest to process. It is the dominant AM material for rapid prototyping, biomedical implants, and consumer products. Thermoplastics like ABS, PLA and nylon are widely used due to their high printability and recyclability. High-performance materials like PEEK and PEI

(ULTEM) are used for aerospace and medical applications due to their high thermal resistance and biocompatibility.

Metal-based AM materials like stainless steel, titanium alloy, Inconel and cobalt have revolutionized industries that require superior strength, corrosion resistance and fatigue resistance. The use of metallic materials processed by an AM process in the automotive, aerospace, and biomedical fields highlights their potential to replace or complement conventional subtractive manufacturing processes. Ceramic materials are challenging to process due to their brittleness and high sintering temperature. Their applications are increasing considerably in biomedical implants and high-strength applications.

The development of hybrid material for AM introduces the option to develop new material with tailored mechanical and functional properties. Further research is expected to find new advanced material for AM to focus on multi-material printing, recyclable, sustainable, and AI-driven material, and *in-situ* process monitoring.

REFERENCES

Agrawal, R. (2021). Sustainable material selection for additive manufacturing technologies: A critical analysis of rank reversal approach. *Journal of Cleaner Production*, *296*, 126500. https://doi.org/10.1016/j.jclepro.2021.126500

Ahuja, B., Karg, M., & Schmidt, M. (2015). Additive manufacturing in production: Challenges and opportunities. *Laser 3D Manufacturing II*, *9353*, 11–20. https://doi.org/10.1117/12.2082521

Alghamdy, M., Ahmad, R., & Alsayyed, B. (2019). Material selection methodology for additive manufacturing applications. *Procedia CIRP*, *84*, 486–490. https://doi.org/10.1016/j.procir.2019.04.296

Boydston, A. J., Cui, J., Lee, C. U., Lynde, B. E., & Schilling, C. A. (2020). 100th anniversary of macromolecular science viewpoint: Integrating chemistry and engineering to enable additive manufacturing with high-performance polymers. *ACS Macro Letters*, *9(8)*, 1119–1129. https://doi.org/10.1021/acsmacrolett.0c00398

Chatterjee, S., Dutta, P., & Banerjee, S. (2023). Sustainability in additive manufacturing: Life cycle and environmental assessment. *Resources, Conservation and Recycling Advances*, *19*, 200197. https://doi.org/10.1016/j.rcradv.2023.200197

DebRoy, T., Wei, H. L., Zuback, J. S., Mukherjee, T., Elmer, J. W., Milewski, J. O., Beese, A. M., Wilson-Heid, A., De, A., & Zhang, W. (2018). Additive manufacturing of metallic components – Process, structure and properties. *Progress in Materials Science*, *92*, 112–224. https://doi.org/10.1016/j.pmatsci.2017.10.001

Deckers, J., Vleugels, J., & Kruth, J. P. (2014). Additive manufacturing of ceramics: A review. *Journal of Ceramic Science and Technology*, *5(4)*, 245–260. https://doi.org/10.4416/JCST2014-00032

Ford, S., & Despeisse, M. (2016). Additive manufacturing and sustainability: An exploratory study of the advantages and challenges. *Journal of Cleaner Production*, *137*, 1573–1587. https://doi.org/10.1016/j.jclepro.2016.04.150

García-Hernández, C., García-Cabezón, C., González-Diez, F., Ampudia, M., Juanes-Gusano, D., Rodriguez-Cabello, J. C., & Martín-Pedrosa, F. (2025). Effect of processing on microstructure, mechanical properties, corrosion and biocompatibility of additive manufacturing Ti-6Al-4V orthopaedic implants. *Scientific Reports*, *15*, 14087. https://doi.org/10.1038/s41598-025-98349-6

Hassan, M. H., Omar, A. M., Daskalakis, E., Hou, Y., Huang, B., Strashnov, I., Askin, S., Iqbal, H., Ferreira, F. C., & Bártolo, P. (2020). The potential of polyethylene terephthalate glycol as biomaterial for bone tissue engineering. *Polymers, 12(12)*, 3045. https://doi.org/10.3390/polym12123045

Huang, S. H., Liu, P., Mokasdar, A., & Hou, L. (2013). Additive manufacturing and its societal impact: A literature review. *International Journal of Advanced Manufacturing Technology, 67(5)*, 1191–1203. https://doi.org/10.1007/s00170-012-4558-5

Jiang, J., Xu, X., & Stringer, J. (2018). Support structures for additive manufacturing: A review. *Journal of Manufacturing and Materials Processing, 2(4)*, 64. https://doi.org/10.3390/jmmp2040064

Kafi, A., Wu, H., Langston, J., Atak, O., Kim, H., Kim, S., Foster, M., Hofmann, D., & Koo, J. H. (2020). Evaluation of additively manufactured ultraperformance polymers to use as thermal protection systems for spacecraft. *Journal of Applied Polymer Science, 137(37)*, 49117. https://doi.org/10.1002/app.49117

Kawalkar, R., Dubey, H. K., & Lokhande, S. P. (2022). A review for advancements in standardization for additive manufacturing. *Materials Today: Proceedings, 50*, 1983–1990. https://doi.org/10.1016/j.matpr.2021.09.417

Kumar, S. A., & Prasad, R. V. S. (2021). Basic principles of additive manufacturing: Different additive manufacturing technologies. In *Additive Manufacturing* (pp. 17–35). https://doi.org/10.1038/s41598-025-98349-6

Liu, C., Tian, W., & Kan, C. (2022). When AI meets additive manufacturing: Challenges and emerging opportunities for human-centered products development. *Journal of Manufacturing Systems, 64*, 648–656. https://doi.org/10.1016/j.jmsy.2022.05.010

Negi, S., Dhiman, S., & Sharma, R. K. (2013). Basics, applications and future of additive manufacturing technologies: A review. *Journal of Manufacturing Technology Research, 5(1–2)*, 75. www.novapublishers.com/catalog/product_info.php?products_id=41415

Ngo, T. D., Kashani, A., Imbalzano, G., Nguyen, K. T., & Hui, D. (2018). Additive manufacturing (3D printing): A review of materials, methods, applications and challenges. *Composites Part B: Engineering, 143*, 172–196. https://doi.org/10.1016/j.compositesb.2018.02.012

Qian, M. A., Xu, W., Brandt, M., & Tang, H. P. (2016). Additive manufacturing and postprocessing of Ti-6Al-4V for superior mechanical properties. *MRS Bulletin, 41(10)*, 775–784. https://doi.org/10.1557/mrs.2016.211

Rajendran, S., Palani, G., Kanakaraj, A., Shanmugam, V., Veerasimman, A., Gądek, S., & Marimuthu, U. (2023). Metal and polymer-based composites manufactured using additive manufacturing—A brief review. *Polymers, 15(11)*, 2564. https://doi.org/10.3390/polym15112564

Sadaf, M., Bragaglia, M., Slemenik Perše, L., & Nanni, F. (2024). Advancements in metal additive manufacturing: A comprehensive review of material extrusion with highly filled polymers. *Journal of Manufacturing and Materials Processing, 8(1)*, 14. https://doi.org/10.3390/jmmp8010014

Santos, K. W., Costa, K. J. S. G., Gonçalves, I. S., Alves, M., Lauda, D. P., Vasconcellos, L. M. R., Araujo, A. H., Seabra, M. P., & Oliveira, I. R. (2024). Influence of manufacturing parameters on bioactive glass 45S5: Structural analysis and applications in bone tissue engineering. *Ceramics International, 50(23)*, 51043–51054. https://doi.org/10.1016/j.ceramint.2024.06.171

Sing, S. L., An, J., Yeong, W. Y., & Wiria, F. E. (2016). Laser and electron-beam powder-bed additive manufacturing of metallic implants: A review on processes, materials and designs. *Journal of Orthopaedic Research, 34(3)*, 369–385. https://doi.org/10.1002/jor.23048

Singh, G., Singh, S., Prakash, C., Kumar, R., Kumar, R., & Ramakrishna, S. (2020). Characterization of three-dimensional printed thermal-stimulus polylactic acid–hydroxyapatite-based shape memory scaffolds. *Polymer Composites*, *41(9)*, 3871–3891. https://doi.org/10.1002/pc.25638

Tan, L. J., Zhu, W., & Zhou, K. (2020). Recent progress on polymer materials for additive manufacturing. *Advanced Functional Materials*, *30(43)*, 2003062. https://doi.org/10.1002/adfm.202003062

Tarafder, S., & Bose, S. (2013). Polycaprolactone-coated 3D printed tricalcium phosphate scaffolds for bone tissue engineering. *Journal of Biomedical Materials Research Part A*, *101(10)*, 3006–3015. https://doi.org/10.1002/jbm.a.34595

Tofail, S. A., Koumoulos, E. P., Bandyopadhyay, A., Bose, S., O'Donoghue, L., & Charitidis, C. (2018). Additive manufacturing: Scientific and technological challenges, market uptake and opportunities. *Materials Today*, *21(1)*, 22–37. https://doi.org/10.1016/j.mattod.2017.07.006

Tshephe, T. S., Akinwamide, S. O., Olevsky, E., & Olubambi, P. A. (2022). Additive manufacturing of titanium-based alloys: A review of methods, properties, challenges, and prospects. *Heliyon*, *8(3)*, e08941. https://doi.org/10.1016/j.heliyon.2022.e08941

Vidakis, N., Maniadi, A., Petousis, M., Vamvakaki, M., Kenanakis, G., & Koudoumas, E. (2020). Mechanical and electrical properties investigation of 3D-printed acrylonitrile–butadiene–styrene graphene and carbon nanocomposites. *Journal of Materials Engineering and Performance*, *29(3)*, 1909–1918. https://link.springer.com/article/10.1007/s11665-020-04689-x

Wong, K. V., & Hernandez, A. (2012). A review of additive manufacturing. *ISRN Mechanical Engineering*, *2012*, 208760. https://doi.org/10.5402/2012/208760

Zocca, A., Colombo, P., Gomes, C. M., & Günster, J. (2015). Additive manufacturing of ceramics: Issues, potentialities, and opportunities. *Journal of the American Ceramic Society*, *98(7)*, 1983–2001. https://doi.org/10.1111/jace.13700

2 Fundamentals of Additive Manufacturing Techniques: Polymer, Metal, and Ceramics

Sulaim Rizvi, Shahazad Ali, Anant Prakash Agrawal, and Kashif Irshad

2.1 INTRODUCTION

3D printing develops objects in 3D space by adding layers, one by one, from a sliced digital 3D model. This method makes it possible to fabricate intricate forms that are usually challenging or expensive to manufacture using conventional manufacturing methods (Berdine et al., 2019; Monteiro et al., 2023). 3D printing began as a tool to create prototypes quickly, but has emerged in healthcare, aerospace, automotive, and consumer product industries (Jakus et al., 2015a; Jakus et al., 2015b; Lee et al., 2010). Technology has stimulated foreign interest and financial investment because it enables the manufacturing of intricate products with design adaptations that require reduced materials (D'Alessio & Christensen, 2019).

Charles Hull developed the first 3D printing technology, stereolithography (SLA), in 1984. Ultraviolet (UV) light is the fundamental technology used in SLA that cures the resin liquid using layer-over-layer subsequent hardening to develop three-dimensional (3D) solid components. The fused deposition modeling (FDM) and selective laser sintering (SLS) technologies, which were introduced in the 1990s, extended the potential of 3D printing significantly, and allowed the use of advanced materials. Such inventions established the foundation for the commercialization of the technology in the 2000s, when it started influencing various fields and causing new manufacturing processes to be developed (Horvath, 2014; Jakus, 2019).

Through 3D printing, the quick prototyping and customization of the product has become very user-friendly. This has resulted in great advantages in most industries. In the medical field, it has revolutionized the design of medical implants, prosthetics, and surgical devices. Lightweight aerospace manufacturers create complex, lightweight components that use less fuel and perform better. In the vehicle industry, 3D printing is adopted to develop new components and print tailor-made ones, such as the custom replacement of old vehicles (Hitzler et al., 2018). To achieve more technological advancement in the future, one will need to integrate

 DOI: 10.1201/9781003643531-2

3D printing that boosts innovation and reduces production cost and waste (Horvath & Cameron, 2018).

2.2 OVERVIEW OF 3D PRINTING TECHNOLOGIES

2.2.1 Stereolithography

Industry has widely adopted stereolithography as an established 3D printing technology (Huang et al., 2020). This sophisticated manufacturing method applies an accurate laser beam of UV light to cure resin layers one by one during the printing process. The software analyzes a digital CAD model and divides it into thin horizontal cross-sectional layers that usually range between 25 and 100 microns in thickness. Each resin cross-section pattern accepts the UV laser beam, which uses galvanometer mirrors to guide the beam onto the surface, while activating photopolymerization for instant solidification, as shown in Figure 2.1. SLA technology stands apart because of its sub-micron resolution level and its ability to produce extremely precise surfaces with ± 0.1 mm accurate dimensions, thus making this process best suited for dental models and jewelry prototypes, alongside microfluidic devices. The main drawbacks of SLA technology include its material restrictions to acrylates and epoxy-based photopolymers, alongside the size limitations of print outputs, and reduced production speed, alongside necessary support service steps for final material properties (Kafle et al., 2021; Schmidleithner & Kalaskar, 2018; Xu et al., 2020).

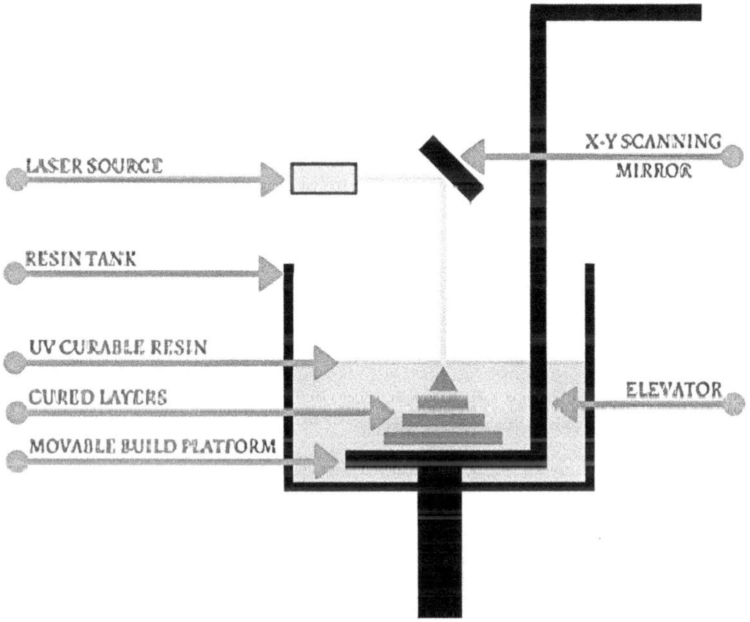

FIGURE 2.1 Stereolithography.

Source: manufactur3dmag.com.

2.2.2 Selective Laser Sintering

Selective laser sintering is an outstanding AM method that uses 50–200 W CO_2 lasers to merge thermoplastic along with metal and ceramic powder particles near their melting temperatures (Kruth et al., 2003). A recoating blade is used to provide a 50–100 micron layer of powder on the platform at the beginning of the intricate SLS process. The laser directs its beam to precisely follow the dimensional profile, and thus enables particle fusion by molecular adhesion instead of complete liquefaction, as shown in Figure 2.2. The main benefit of SLS technology is its natural scaffold formation from unfused powder that obviates the need for support structures, thus facilitating the production of intricate components with internal features and complex lattice structures. The SLS process generates rugged parts with properties similar to injection-molded materials, but it faces limits such as surface roughness at 7–10 μm, and requires cool-down phases between builds, alongside possible heat-related deformations (Fina et al., 2017; Schmid et al., 2015).

2.2.3 Fused Deposition Modeling

Fused deposition modeling printers melt thermoplastic filament before discharging it from a heated nozzle for consecutive layer creation to fabricate objects (Mwema et al., 2020). The computer-generated model gets split into multiple thin layers by the software, while a printing device uses the melted strand to construct the object, starting from its base layers until completion, as shown in Figure 2.3. The low-cost and user-friendly nature of FDM made it a commonly used technique. The resolution, as well as the surface finish capabilities of FDM 3D printing, is inferior to that provided by other 3D printing technologies. FDM operates for the production of prototypes and with hobby crafts, along with industrial support applications (Kristiawan et al., 2021; Mohamed et al., 2015; Rajan et al., 2022).

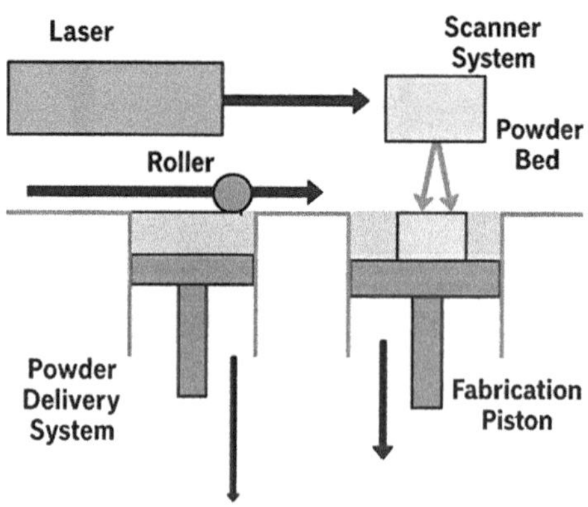

FIGURE 2.2 Selective laser sintering.

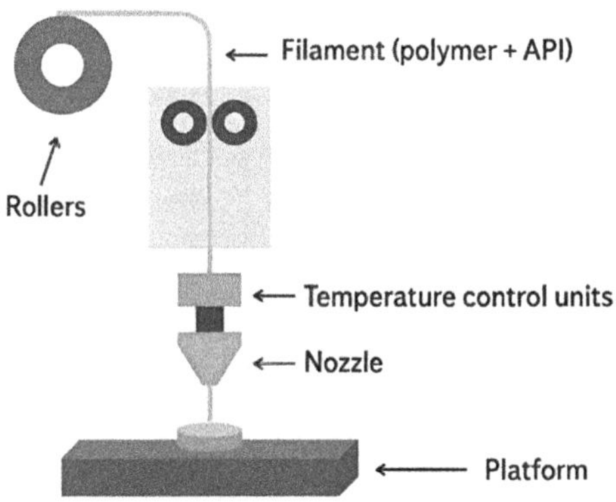

FIGURE 2.3 Fused deposition modeling.

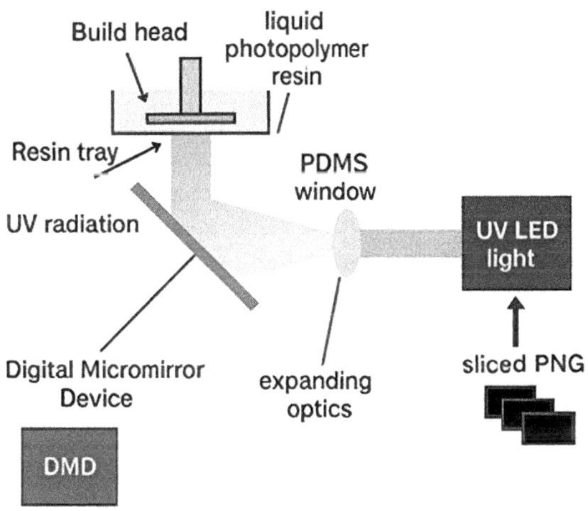

FIGURE 2.4 Digital light processing.

2.2.4 Digital Light Processing

Digital light processing (DLP) printing operates in a manner similar to SLA, using photopolymer resins and a digital light projector as the light source. It builds objects by projecting images of each cross-section, curing resin layers one at a time. Thanks to its high resolution and speed, DLP is ideal for applications that demand precision, such as dental model production and jewelry manufacturing. This process is not only limited to light-curable resins, but also requires more post-processing (Kadry et al., 2019; Krkobabić et al., 2020). A schematic diagram of DLP is shown Figure 2.4.

2.2.5 Electron Beam Melting

Electron beam melting (EBM) uses a high-energy electron beam in a vacuum to melt layers of metal powder. The vacuum is required to reduce the chances of

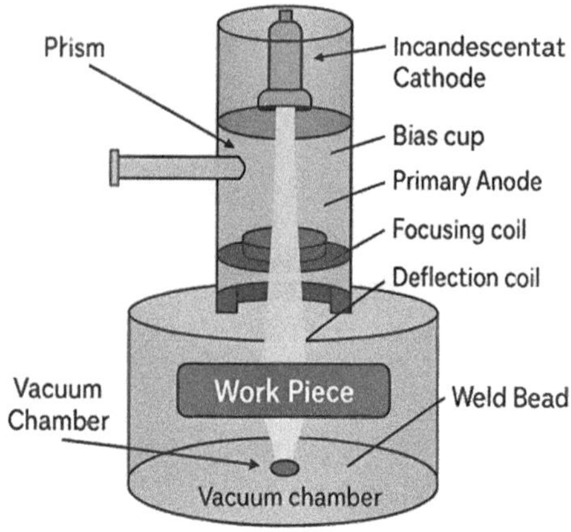

FIGURE 2.5 Electron beam melting.

contamination and oxidation, and also to enhance melting precision (Konda Gokuldoss et al., 2017; Kolamroudi et al., 2021). EBM is specifically efficient for making complex, high-performance elements, having major applications in aerospace and medical equipment. The parts manufactured by this process demonstrate good properties, strength, and stability of the structure. Limitations that accompany this process include the expensive set up and high initial cost, expensive operation cost, complicated post-processing, and the need for skilled operators (Bikas et al., 2016; Liu et al., 2020; Uçak et al., 2022). A schematic diagram of EBM is shown Figure 2.5.

2.3 APPLICATIONS IN VARIOUS FIELDS

2.3.1 Bioprinting Applications

Bioprinting corresponds to the most innovative form of 3D printing, emphasizing the printing of organs and tissues. It works by depositing layers of living cells and biomaterials to build highly precise structures that closely replicate natural biological systems. Traditional methods lack the development of fully functional and vascularized tissues and body parts (Fang et al., 2024; Mandrycky et al., 2016). Nevertheless, the techniques of bioprinting have significantly enhanced the structured organization of a number of cells, extracellular matrix (ECM) molecules, and growth factors with enhanced vascularization and the functionality of engineered tissue constructs (Gauvin et al., 2012; Wei et al., 2024). Such constructs exhibit potential in the area of tissue implants, disease modeling, drug screening, and drug discovery (Jeong et al., 2012). Commonly used bioprocess techniques include inkjet, extrusion, laser-based processes, stereolithography, and acoustic methods. Each of these methods hold inherent advantages for specific uses. Bioprinting is used for the construction of artificial body parts such as cartilage, skin, liver, lung, neural tissue, and the pancreas with a specific microarchitecture (Binder et al., 2010). Bioprinting developments in the recent past have concentrated on designing vascular networks in tissue constructs both indirectly and directly. The advances have tremendously increased the avenue of tissue engineering, and have advanced the prospects of regenerative medicine (Khalil & Sun, 2009).

2.3.2 Dental Applications

Additive manufacturing is changing the face of dentistry by allowing the production of dental implants, surgical guides, and anatomical models through a digital model based on CAD software through 3D printing. This enables very accurate and customized dental care that will enhance patient satisfaction and reduce the time spent in the procedures. It is employed to make orthodontic, prosthodontic, and endodontic instruments like custom braces, aligners, crowns, bridges, and root canal guides with great precision (Shaikh et al., 2021). In dentistry, 3D-printed models of hard and soft tissues provide students with realistic practice models, which seem to be more effective in training and improving their skills (Javaid & Haleem, 2019). The integration of 3D printing and the latest imaging tools, such as dental cone beam computed tomography, and magnetic resonance imaging offer insightful images of patient-specific anatomical features. This integration is very effective in planning and performing complicated dental procedures. Altogether, 3D printing evolves the areas of personalization, accuracy, and efficiency in dental treatments, as well as dental education (Alharbi et al., 2017; Dawood et al., 2015).

2.3.3 Aerospace Applications

Additive manufacturing has transformed the aerospace sector by assisting in the production of complex, lightweight components that enhance efficiency and bring about cost savings (Fu et al., 2022; Karkun & Dharmalingam, 2022; Joshi & Sheikh, 2015). This technology is employed for the construction of both structural and non-structural components, such as fuel nozzles, brackets, and ducting systems that lead to huge fuel consumption savings and low emissions. The Federal Aviation Administration (FAA) approved the first 3D-printed commercial airplane jet engine part from General Electric (GE) in 2015. The inlet temperature sensors were mounted in the inlet of the compressor of a jet engine. GE Aviation and Boeing are jointly retrofitting the GE90–94B jet engines with more than 400 3D-printed components. GE's LEAP jet engine, with 19 3D-printed fuel nozzles, is already on a test flight with the aim of implementing it in Boeing 737MAX and Airbus A320 neoplanes.

Super Hornet planes incorporate around 100 AM components for air-cooler ducts. Components for the Bell 429 have likewise been manufactured with laser-sintering technology with the vision of utilizing more laser-sintered components in other helicopters (Khajavi et al., 2014). Additive manufacturing has also been used for the production of a wing spar on the Lockheed Martin F-35 and a GKN Aerospace-produced wing leading edge on the Dassault Falcon 5X. Dozens of titanium alloy brackets have already been installed in the solar-powered Juno spaceship by Lockheed Martin using EBM technology (Tadjdeh, 2014). Boeing is also applying 3D printing in the production of plastic interior components using primarily ULTEM and nylon for prototypes and test coupons.

2.3.4 Space Technology Applications

Space technology has shown a lot of potential with 3D printing, especially for use on the International Space Station (Hall Curt, 2024). Long-duration missions are supported by the capability to produce tools and parts in space whenever required, which reduces the need for large spare components (Zocca et al., 2022). The possibility of building houses and other buildings on other planets using local resources, like soil from the moon or Mars, is also being investigated by researchers. Additionally, in space missions, 3D printing can be extremely helpful in medical emergencies (Cubo-Mateo & Gelinsky, 2021; Karlin & Bauer, 2019).

2.3.5 Food Technology Applications

Additive manufacturing gives food producers and chefs an opportunity to design individual meals that suit the individual requirements of each person in terms of their diet, and to experiment with new culinary options (Kim et al., 2020). There are applications that could dramatically alter food production and consumption, including high-end food and industrial production. The application of 3D printing to the construction of complex food designs with a successful implementation of the whole dining experience is demonstrated in several case studies (Godoi et al., 2016; Liu et al., 2017; Nachal et al., 2019; Sun et al., 2015; Taneja et al., 2022).

2.3.6 Automotive Applications

Additive manufacturing technology is impacting the automotive industry in various ways, such as quick prototyping and making custom car parts (Nichols, 2019). Through technology, producers can test new designs quickly and efficiently, cutting down waste and scrap, as well as development time and costs (Nascimento et al., 2022). Moreover, 3D printing simplifies the production of customized components of old vehicles. This means that old cars can be restored, as businesses such as Ford and BMW have used 3D printing to enhance the efficiency in their supply chain, and streamline their production processes (Raina et al., 2021; Shahrubudin et al., 2019).

2.3.7 Education Sector

The use of 3D printing equipment in educational institutes has proved to have very strong pedagogical benefits in various subjects. The application of 3D printing helps to create real-life models of abstract things, thus increasing student understanding and the memorization of complicated material. Learning institutions that use this technology have recorded quantifiable gains in student engagement rates and conceptual knowledge, especially in science, technology, engineering and mathematics (STEM) fields when theoretical constructs would inherently be difficult to visualize.

The downward trend of the prices of 3D printing equipment has increased the availability of the technology, making it possible to apply it in more educational environments and in different socioeconomic backgrounds (Assante et al., 2020; Novak, 2022).

2.4 TECHNOLOGICAL ADVANCEMENTS AND INNOVATIONS

2.4.1 Materials Development

The 3D printing of materials has resulted in long-lasting and biocompatible materials that are produced by materials science. These materials are essential in the industrial or medical world, since they enable the manufacture of reliable and working parts (Praveena et al., 2022). Specific applications, such as bioinks and conductive inks, have more possibilities due to the arrival of new materials. There is also increased attention on sustainable and environmentally friendly printing materials, since their effect on manufacturing is low (Jamróz et al., 2018).

2.4.2 Software and Digital Integration

The development of CAD software and its adaptation to 3D printing have significantly enhanced the process of printing. State-of-the-art software programs automate the production process, facilitate, to a greater extent, more accurate and faster workflow, reduce errors, and enhance the quality of prints. The entire process of production is streamlined through uninterrupted coordination via the computerized synthesis of designers, engineers, and manufacturers (Do, 2017; Junk & Matt, 2015; Lee et al., 2017).

2.4.3 Precision and Customization

The actual changes in 3D printing technology in recent times have resulted in a remarkable breakthrough in the accuracy of printing, as well as customization. The 3D-printed models in dental education that represent hard and soft tissues effectively give students a realistic experience. This enables them to grow and sharpen their skills better to create an improved training process. Several case studies show how these tailor-made solutions can enhance better patient outcomes, and address particular user requirements (Lee et al., 2017; Mahmood et al., 2020).

2.4.4 Challenges and Limitations

Although the benefits of 3D printing are numerous, it has its shortcomings as well. Technical constraints include issues like material quality, and printing speed and resolution. In addition to this, the expensive nature of investment, the expensive nature of accessories, and high operating costs limit the application of these 3D printing technologies to smaller industries in developing countries. One must also make sure that these technologies are applied responsibly in various fields, such as in the printing of organs and defense (Goh et al., 2021; Mostafaei et al., 2021).

2.5 UNTAPPED AREAS FOR 3D PRINTING

Even though 3D printing has achieved significant breakthroughs in certain areas, opportunities for the use of this technology in other areas exist and remain unexplored. There are prospects in the architectural, textile, and food industries, in which 3D printing can be used. It has the potential of offering something new to solve urgent issues. The study of these areas may lead to other applications.

3D printing could change precision farming in agriculture because it would facilitate the production of individualized tools and apparatus to meet the needs of a particular farm. Particular nozzles for irrigation systems, agricultural surveillance drones and special parts, and accurate planting are tools that are aimed at maximizing resource utilization, growing crop volumes, and minimizing the environmental influence. 3D printing can also be applied to the production of tailor-made sensors and monitoring devices that measure plant health, nutrient content, and moisture in the soil. This is a technology that provides farmers with real-time information, allowing them to make smarter decisions regarding pests, fertilization and watering. Another innovative method is the development of biodegradable seed pods that cover young plants and protect them from pests and harsh weather. Such pods disintegrate naturally in the soil and enhance germination.

In the textile industry, 3D printing can revolutionize the manufacture of tailored clothing products. Scanning a person's body, tailor-made garments could be produced that are customized in size. This would minimize waste as the requirement for overproduction and leftover inventory will be reduced. In addition, 3D printing facilitates the manufacture of complex fabric structures, which are hard to produce by standard methods of textile manufacture. This involves fabrics with inbuilt sensors, materials that adapt the properties depending on the condition of the environment, and fabrics that have a specialty aesthetic design. The use of sustainable and recycled materials in fabrication can also contribute to a minimization of the fashion industry's deterioration of the environment. With the flexibility to design clothes on demand, designers can reduce both waste and the carbon footprint due to transporting and storing larger inventories.

It is a technology that enables faster design cycles and the manufacture of customized components that enhance the appearance and functionality of building, and reduce development time. On-site 3D printing strikingly reduces the cost of labor and speeds up construction. Large-scale printers can print walls, columns, and entire structures with cement-based materials or composites. In addition to building, 3D printing will have benefits in other industries. In disaster relief and affordable housing, it can assist in constructing transitional or permanent shelters quicker.

3D printing has facilitated rapid prototyping and the personalization of products in a positive way, impacting many sectors. It has transformed the manufacture of medical implants in healthcare, tooth prostheses, and surgical instruments. It is used by aerospace manufacturers to develop light complexes, and parts that make a car more fuel-efficient and performance-driven. The motor industry also uses 3D printing to create new components and make customized parts of vintage cars for enthusiasts. (Agrawal et al., 2023; Raja et al., 2022; Srivastava et al., 2025).

2.6 CONCLUSION

The review outlined the ability of AM to revolutionize many industries through innovation, a reduction in the cost of production, and reducing the wastage of materials. Many uses in healthcare, aerospace, the automobile industry, and education demonstrate the versatility of the technology, and its high potential to continue its development. Levels of development in the sphere of materials science, methods, and cross-disciplinary teams are to undergo a further increase in the possibilities of 3D printing technology, making it one of the cornerstones of the future development of industries. The future of AM is optimistic. This has huge potential for further scientific research and application.

REFERENCES

Agrawal, A. P., Kumar, V., Kumar, J., Paramasivam, P., Dhanasekaran, S., & Prasad, L. (2023). An investigation of combined effect of infill pattern, density, and layer thickness on mechanical properties of 3D printed ABS by fused filament fabrication. *Heliyon*, 9(6), 1–12. https://doi.org/10.1016/j.heliyon.2023.e16531

Alharbi, N., Wismeijer, D., & Osman, R. B. (2017). Additive manufacturing techniques in prosthodontics: Where do we currently stand? A critical review. *International Journal of Prosthodontics*, 30(5), 1–11. DOI: 10.11607/ijp.5079.

Assante, D., Cennamo, G. M., & Placidi, L. (2020, April). 3D printing in education: An European perspective. In *2020 IEEE global engineering education conference (EDUCON)* (pp. 1133–1138). IEEE. https://doi.org/10.1109/EDUCON45650.2020.9125311

Berdine, G. G., DiPaola, M., & Weinberg, M. (2019). Economic and regulatory perspectives on additive manufacturing. In *3D Printing in Orthopaedic Surgery* (pp. 41–48). Elsevier. https://doi.org/10.1016/B978-0-323-58118-9.00004-X

Bikas, H., Stavropoulos, P., & Chryssolouris, G. (2016). Additive manufacturing methods and modelling approaches: A critical review. *The International Journal of Advanced Manufacturing Technology*, 83, 389–405. https://doi.org/10.1007/s00170-015-7576-2

Binder, K. W., Zhao, W., Aboushwareb, T., Dice, D., Atala, A., & Yoo, J. J. (2010). In situ bioprinting of the skin for burns. *Journal of the American College of Surgeons*, 211(3), S76. https://doi.org/10.1016/j.jamcollsurg.2010.06.198

Cubo-Mateo, N., & Gelinsky, M. (2021). Wound and skin healing in space: The 3D bioprinting perspective. *Frontiers in Bioengineering and Biotechnology*, 9, 720217. https://doi.org/10.3389/fbioe.2021.720217

D'Alessio, J., & Christensen, A. (2019). 3D printing for commercial orthopedic applications: Advances and challenges. In Dipaola, M. & Wodajo, F. M. (Eds.). *3D Printing in Orthopaedic Surgery* (pp. 65–83). https://doi.org/10.1016/B978-0-323-58118-9.00007-5

Dawood, A., Marti, B. M., Sauret-Jackson, V., & Darwood, A. (2015). 3D printing in dentistry. *British Dental Journal*, 219(11), 521–529. https://doi.org/10.1038/sj.bdj.2015.914

Do, N. (2017). Integration of design and manufacturing data to support personal manufacturing based on 3D printing services. *The International Journal of Advanced Manufacturing Technology*, 90(9), 3761–3773. https://doi.org/10.1007/s00170-016-9688-8

Fang, W., Yu, Z., Gao, G., Yang, M., Du, X., Wang, Y., & Fu, Q. (2024). Light-based 3D bioprinting technology applied to repair and regeneration of different tissues: A rational proposal for biomedical applications. *Materials Today Bio*, 27, 101135. https://doi.org/10.1016/j.mtbio.2024.101135

Fina, F., Goyanes, A., Gaisford, S., & Basit, A. W. (2017). Selective laser sintering (SLS) 3D printing of medicines. *International Journal of Pharmaceutics*, 529(1–2), 285–293. https://doi.org/10.1016/j.ijpharm.2017.06.082

Fu, X., Lin, Y., Yue, X. J., XunMa, Hur, B., & Yue, X. Z. (2022, May). A review of additive manufacturing (3D printing) in aerospace: Technology, materials, applications, and challenges. In *Mobile Wireless Middleware, Operating Systems and Applications: 10th International Conference on Mobile Wireless Middleware, Operating Systems and Applications (MOBILWARE 2021)* (pp. 73–98). Springer International Publishing. https://doi.org/10.1007/978-3-030-98671-1_6

Gauvin, R., Chen, Y. C., Lee, J. W., Soman, P., Zorlutuna, P., Nichol, J. W., ... & Khademhosseini, A. (2012). Microfabrication of complex porous tissue engineering scaffolds using 3D projection stereolithography. *Biomaterials*, 33(15), 3824–3834. https://doi.org/10.1016/j.biomaterials.2012.01.048

Godoi, F. C., Prakash, S., & Bhandari, B. R. (2016). 3d printing technologies applied for food design: Status and prospects. *Journal of Food Engineering*, 179, 44–54. https://doi.org/10.1016/j.jfoodeng.2016.01.025

Goh, G. D., Sing, S. L., & Yeong, W. Y. (2021). A review on machine learning in 3D printing: Applications, potential, and challenges. *Artificial Intelligence Review*, 54(1), 63–94. https://doi.org/10.1007/s10462-020-09876-9

Hall Curt. (2024). 3D printing & the future of space exploration, www.cutter.com/article/3d-printing-future-space-exploration.

Hitzler, L., Alifui-Segbaya, F., Williams, P., Heine, B., Heitzmann, M., Hall, W., ... & Öchsner, A. (2018). Additive manufacturing of cobalt-based dental alloys: Analysis of microstructure and physicomechanical properties. *Advances in Materials Science and Engineering*, 2018(1), 8213023. https://doi.org/10.1155/2018/8213023

Horvath, J. (2014). A brief history of 3D printing. In *Mastering 3D Printing* (pp. 3–10). Apress. https://doi.org/10.1007/978-1-4842-0025-4_1

Horvath, J., & Cameron, R. (2018). Mastering 3D printing in the classroom, library, and lab. *Apress*. https://doi.org/10.1007/978-1-4842-3501-0

Huang, J., Qin, Q., & Wang, J. (2020). A review of stereolithography: Processes and systems. *Processes*, 8(9), 1138. https://doi.org/10.3390/pr8091138

Jakus, A. E. (2019). An introduction to 3D printing – past, present, and future promise. In *3D Printing in Orthopaedic Surgery* (pp. 1–15). Elsevier. https://doi.org/10.1016/B978-0-323-58118-9.00001-4

Jakus, A. E., Secor, E. B., Rutz, A. L., Jordan, S. W., Hersam, M. C., & Shah, R. N. (2015a). Three-dimensional printing of high-content graphene scaffolds for electronic and biomedical applications. *ACS Nano*, 9(4), 4636–4648. https://doi.org/10.1021/acsnano.5b01179

Jakus, A. E., Taylor, S. L., Geisendorfer, N. R., Dunand, D. C., & Shah, R. N. (2015b). Metallic architectures from 3D-printed powder-based liquid inks. *Advanced Functional Materials*, 25(45), 6985–6995. https://doi.org/10.1002/adfm.201503921

Jamróz, W., Szafraniec, J., Kurek, M., & Jachowicz, R. (2018). 3D printing in pharmaceutical and medical applications–recent achievements and challenges. *Pharmaceutical Research*, 35, 1–22. https://doi.org/10.1007/s11095-018-2454-x

Javaid, M., & Haleem, A. (2019). Current status and applications of additive manufacturing in dentistry: A literature-based review. *Journal of Oral Biology and Craniofacial Research*, 9(3), 179–185. https://doi.org/10.1016/j.jobcr.2019.04.004

Jeong, J. H., Chan, V., Cha, C., Zorlutuna, P., Dyck, C., Hsia, K. J., ... & Kong, H. (2012). "Living" microvascular stamp for patterning of functional neovessels; orchestrated control of matrix property and geometry. *Advanced Materials*, 24(1), 58–63. https://doi.org/10.1002/adma.201103207

Joshi, S. C., & Sheikh, A. A. (2015). 3D printing in aerospace and its long-term sustainability. *Virtual and Physical Prototyping*, 10(4), 175–185. https://doi.org/10.1080/17452759.2015.1111519

Junk, S., & Matt, R. (2015). New approach to introduction of 3D digital technologies in design education. *Procedia Cirp*, 36, 35–40. https://doi.org/10.1016/j.procir.2015.01.045

Kadry, H., Wadnap, S., Xu, C., & Ahsan, F. (2019). Digital light processing (DLP) 3D-printing technology and photoreactive polymers in fabrication of modified-release tablets. *European Journal of Pharmaceutical Sciences*, 135, 60–67. https://doi.org/10.1016/j.ejps.2019.05.008

Kafle, A., Luis, E., Silwal, R., Pan, H. M., Shrestha, P. L., & Bastola, A. K. (2021). 3D/4D printing of polymers: Fused deposition modelling (FDM), selective laser sintering (SLS), and stereolithography (SLA). *Polymers*, 13(18), 3101. https://doi.org/10.3390/polym13183101

Karkun, M. S., & Dharmalingam, S. (2022). 3D printing technology in aerospace industry – a review. *International Journal of Aviation, Aeronautics, and Aerospace*, 9(2), 4. https://doi.org/10.15394/ijaaa.2022.1708

Karlin, L. I., & Bauer, A. S. (2019). Pediatric spine and extremity. In *3D Printing in Orthopaedic Surgery* (pp. 137–150). Elsevier. https://doi.org/10.1016/B978-0-323-58118-9.00012-9

Khajavi, S. H., Partanen, J., & Holmström, J. (2014). Additive manufacturing in the spare parts supply chain. *Computers in Industry*, 65(1), 50–63. https://doi.org/10.1016/j.compind.2013.07.008

Khalil, S., & Sun, W. (2009). Bioprinting endothelial cells with alginate for 3D tissue constructs. *Journal of Biomechanical Engineering*, 131, 1–8. https://doi.org/10.1115/1.3128729

Kim, M. J., Kim, M. K., & You, Y. S. (2020). Food 3D printing technology and food materials of 3D printing. *Clean Technology*, 26(2), 109–115. https://doi.org/10.7464/ksct.2020.26.2.109

Kolamroudi, M. K., Asmael, M., Ilkan, M., & Kordani, N. (2021). Developments on electron beam melting (EBM) of Ti–6Al–4V: A review. *Transactions of the Indian Institute of Metals*, 74, 783–790. https://doi.org/10.1007/s12666-021-02230-9

Konda Gokuldoss, P., Kolla, S., & Eckert, J. (2017). Additive manufacturing processes: Selective laser melting, electron beam melting and binder jetting – Selection guidelines. *Materials*, 10(6), 672. https://doi.org/10.3390/ma10060672

Kristiawan, R. B., Imaduddin, F., Ariawan, D., Ubaidillah, & Arifin, Z. (2021). A review on the fused deposition modeling (FDM) 3D printing: Filament processing, materials, and printing parameters. *Open Engineering*, 11(1), 639–649. https://doi.org/10.1515/eng-2021-0066

Krkobabić, M., Medarević, D., Pešić, N., Vasiljević, D., Ivković, B., & Ibrić, S. (2020). Digital light processing (DLP) 3D printing of atomoxetine hydrochloride tablets using photoreactive suspensions. *Pharmaceutics*, 12(9), 833. https://doi.org/10.3390/pharmaceutics12090833

Kruth, J. P., Wang, X., Laoui, T., & Froyen, L. (2003). Lasers and materials in selective laser sintering. *Assembly Automation*, 23(4), 357–371 https://doi.org/10.1108/01445150310698652

Lee, C. H., Cook, J. L., Mendelson, A., Moioli, E. K., Yao, H., & Mao, J. J. (2010). Regeneration of the articular surface of the rabbit synovial joint by cell homing: A proof of concept study. *The Lancet*, 376(9739), 440–448. https://doi.org/10.1016/S0140-6736(10)60668-X

Lee, J., Kim, K. E., Bang, S., Noh, I., & Lee, C. (2017). A desktop multi-material 3D bio-printing system with open-source hardware and software. *International Journal of*

Precision Engineering and Manufacturing, 18, 605–612. https://doi.org/10.1007/s12541-017-0072-x

Liu, C., Yan, D., Tan, J., Mai, Z., Cai, Z., Dai, Y., ... & Chen, Z. (2020). Development and experimental validation of a hybrid selective laser melting and CNC milling system. *Additive Manufacturing,* 36, 101550. https://doi.org/10.1016/j.addma.2020.101550

Liu, Z., Zhang, M., Bhandari, B., & Wang, Y. (2017). 3D printing: Printing precision and application in food sector. *Trends in Food Science & Technology*, 69, 83–94. https://doi.org/10.1016/j.tifs.2017.08.018

Mahmood, M. A., Visan, A. I., Ristoscu, C., & Mihailescu, I. N. (2020). Artificial neural network algorithms for 3D printing. *Materials*, 14(1), 163. https://doi.org/10.3390/ma14010163

Mandrycky, C., Wang, Z., Kim, K., & Kim, D. H. (2016). 3D bioprinting for engineering complex tissues. *Biotechnology Advances*, 34(4), 422–434. https://doi.org/10.1016/j.biotechadv.2015.12.011

Mohamed, O. A., Masood, S. H., & Bhowmik, J. L. (2015). Optimization of fused deposition modeling process parameters: A review of current research and future prospects. *Advances in Manufacturing*, 3, 42–53. https://doi.org/10.1007/s40436-014-0097-7

Monteiro, S. A., Scheid, C., Deon, M., & Merib, J. (2023). Fundamentals, recent applications, and perspectives of 3D printing in sample preparation approaches. *Microchemical Journal*, 195, 109385. https://doi.org/10.1016/j.microc.2023.109385

Mostafaei, A., Elliott, A. M., Barnes, J. E., Li, F., Tan, W., Cramer, C. L., ... & Chmielus, M. (2021). Binder jet 3D printing – Process parameters, materials, properties, modeling, and challenges. *Progress in Materials Science,* 119, 100707. https://doi.org/10.1016/j.pmatsci.2020.100707

Mwema, F. M., Akinlabi, E. T., Mwema, F. M., & Akinlabi, E. T. (2020). Basics of fused deposition modelling (FDM). In Mwema, F. M. & Akinlabi, E. T. (Eds.), *Fused Deposition Modeling: Strategies for Quality Enhancement* (pp. 1–15). https://doi.org/10.1007/978-3-030-48259-6_1

Nachal, N., Moses, J. A., Karthik, P., & Anandharamakrishnan, C. (2019). Applications of 3D printing in food processing. *Food Engineering Reviews*, 11(3), 123–141. https://doi.org/10.1007/s12393-019-09199-8

Nascimento, de Mattos., D. L., Mury Nepomuceno, R., Caiado, R. G. G., Maqueira, J. M., Moyano-Fuentes, J., & Garza-Reyes, J. A. (2022). A sustainable circular 3D printing model for recycling metal scrap in the automotive industry. *Journal of Manufacturing Technology Management*, 33(5), 876–892. https://doi.org/10.1108/JMTM-10-2021-0391

Nichols, M. R. (2019). How does the automotive industry benefit from 3D metal printing?. *Metal Powder Report*, 74(5), 257–258. https://doi.org/10.1016/j.mprp.2019.07.002

Novak, E. (2022). *Physical touch in psychoanalytic psychotherapy: Transforming trauma through embodied practice*. Routledge. https://doi.org/10.4324/9781003215745

Praveena, B. A., Lokesh, N., Santhosh, N., Praveena, B. L., & Vignesh, R. (2022). A comprehensive review of emerging additive manufacturing (3D printing technology): Methods, materials, applications, challenges, trends and future potential. *Materials Today: Proceedings*, 52, 1309–1313. https://doi.org/10.1016/j.matpr.2021.04.254

Raina, A., Haq, M. I. U., Javaid, M., Rab, S., & Haleem, A. (2021). 4D printing for automotive industry applications. *Journal of The Institution of Engineers (India): Series D*, 102, 1–9. https://doi.org/10.1007/s40033-021-00284-z

Raja, S., Agrawal, A. P., P Patil, P., Thimothy, P., Capangpangan, R. Y., Singhal, P., & Wotango, M. T. (2022). Optimization of 3D printing process parameters of polylactic acid filament based on the mechanical test. *International Journal of Chemical Engineering*, 2022(1), 5830869. https://doi.org/10.1155/2022/5830869

Rajan, K., Samykano, M., Kadirgama, K., Harun, W. S. W., & Rahman, M. M. (2022). Fused deposition modeling: Process, materials, parameters, properties, and applications. *The International Journal of Advanced Manufacturing Technology*, 120(3), 1531–1570. https://doi.org/10.1007/s00170-022-08860-7

Schmid, M., Amado, A., & Wegener, K. (2015, May). Polymer powders for selective laser sintering (SLS). In *AIP Conference Proceedings* (Vol. 1664, No. 1). AIP Publishing. https://doi.org/10.1063/1.4918516

Schmidleithner, C., & Kalaskar, D. M. (2018). Stereolithography. *IntechOpen*. https://doi.org/10.5772/intechopen.78147

Shahrubudin, N., Lee, T. C., & Ramlan, R. J. P. M. (2019). An overview on 3D printing technology: Technological, materials, and applications. *Procedia Manufacturing*, 35, 1286–1296. https://doi.org/10.1016/j.promfg.2019.06.089

Shaikh, S., Nahar, P., & Ali, H. M. (2021). Current perspectives of 3d printing in dental applications. *Brazilian Dental Science*, 24(3), 1–9. https://doi.org/10.14295/bds.2021.v24i3.2481

Srivastava, P., Arumugam, A. B., Agrawal, A. P., & Ntumba, Z. N. (2025). Development of hemp fiber reinforced PLA composites for sustainable 3D printing: Mechanical and microstructural properties. *Journal of Natural Fibers*, 22(1), 2558214. https://doi.org/10.1080/15440478.2025.2558214

Sun, J., Zhou, W., Huang, D., Fuh, J. Y., & Hong, G. S. (2015). An overview of 3D printing technologies for food fabrication. *Food and Bioprocess Technology*, 8, 1605–1615. https://doi.org/10.1007/s11947-015-1528-6

Tadjdeh, Y. (2014). 3D printing promises to revolutionize defense, aerospace industries. *National Defense*, 98(724), 20–23. www.jstor.org/stable/27020306

Taneja, A., Sharma, R., Ayush, K., Sharma, A., Mousavi Khaneghah, A., Regenstein, J. M., ... & Sharma, S. (2022). Innovations and applications of 3-D printing in food sector. *International Journal of Food Science and Technology*, 57(6), 3326–3332. https://doi.org/10.1111/ijfs.15691

Uçak, N., Çiçek, A., & Aslantas, K. (2022). Machinability of 3D printed metallic materials fabricated by selective laser melting and electron beam melting: A review. *Journal of Manufacturing Processes*, 80, 414–457. https://doi.org/10.1016/j.jmapro.2022.06.023

Wei, Q., An, Y., Zhao, X., Li, M., & Zhang, J. (2024). Three-dimensional bioprinting of tissue-engineered skin: Biomaterials, fabrication techniques, challenging difficulties, and future directions: A review. *International Journal of Biological Macromolecules*, 266, 131281. https://doi.org/10.1016/j.ijbiomac.2024.131281

Xu, X., Robles-Martinez, P., Madla, C. M., Joubert, F., Goyanes, A., Basit, A. W., & Gaisford, S. (2020). Stereolithography (SLA) 3D printing of an antihypertensive polyprintlet: Case study of an unexpected photopolymer-drug reaction. *Additive Manufacturing*, 33, 101071. https://doi.org/10.1016/j.addma.2020.101071

Zocca, A., Wilbig, J., Waske, A., Günster, J., Widjaja, M. P., Neumann, C., ... & Tian, X. (2022). Challenges in the technology development for additive manufacturing in space. *Chinese Journal of Mechanical Engineering: Additive Manufacturing Frontiers*, 1(1), 100018. https://doi.org/10.1016/j.cjmeam.2022.100018

3 Characterization Techniques for Additive Manufactured Materials

Vinay Shah, Satbir Singh Sehgal, Jasgurpreet Singh Chohan, and Manpreet Singh

3.1 INTRODUCTION

Additive manufacturing (AM) is a process that is also known by the name of 3D printing in industry. 3D printing is considered to be a revolutionary method as it is used to develop lighter and stronger parts. Additive manufacturing has transformed modern manufacturing as it makes it easier to make complex parts with higher accuracy and less wastage of material in the same time.

Wastage of material in the case of AM and rapid prototyping techniques depends highly on the process parameter that has been chosen to make a product as it impacts on its properties and characteristics. Understanding the impact of these parameters, it is important to optimize the performance and reliability of the AM-fabricated part, as this is directly proportional to the process parameters. Traditional subtraction manufacturing techniques require material removal from a solid block to form parts.

Additive manufacturing creates an object layer by layer by exacting data from a digital model. It provides an advantage over the conventional method, where solid material is converted into a final product by subtracting or removing material from an object. This cutting-edge technology continues to transform multiple sectors, including aerospace and automotive manufacturing, and healthcare services and consumer goods.

Additive manufacturing is the integration of various techniques, such as fused deposition modeling (FDM), selective laser sintering (SLS), stereolithography (SLA), and direct metal laser sintering (DMLS). Different AM methods offer specific benefits based on the required materials and precision, as well as the intended applications. Additive manufacturing allows designers and engineers to develop prototypes quickly, while minimizing waste and producing intricate geometries, which traditional manufacturing methods could not achieve.

Additive manufacturing is considered to date back to 1980 when Dr. Hideo Kodama came up with the rapid prototyping technique in Japan. But it was really Charles W. Hull who took things to the next level by creating the first stereolithography printer in 1986. This officially launched the era of commercial 3D printing (Fayazfar et al., 2018). Thereafter, many advancements took place in the computing processes and material that have enhanced the AM process, making it more suitable

DOI: 10.1201/9781003643531-3

and effective. In the 1990s and early 2000s, AM technologies began to branch out into various markets, finding application in everything from dental prosthetics to custom manufacturing tools. The certain rise in the demand for open platform and desktop availability, as well as a pocket-friendly budget, makes AM more popular and aligned with various fields. Nowadays, industries are harnessing AM for large-scale production, with companies weaving it into their supply chains to boost efficiency and sustainability.

The application and future demand of this technique is very high, as its applications range from the basics to very precise components. Take an example from the aerospace industry, where parts are made of very lightweight material, thereby reducing fuel consumption, efficiency, weight ratio, and structural integrity. In the automobile sector, AM is used to produce a customized product and improve the performance of vehicles in terms of mileage and reducing the overall pricing. In the field of biomedical science or the healthcare industry, it plays a very significant role in making implants, prosthetics, and bioprinted tissues.

In futuristic approaches, this technique leads to remarkable inventions. Emerging trends include various types of multi-material printing, where the material properties can be enhanced by including reinforced material in the base filament of AM, which permits the creation of complex geometries and shapes that can easily be developed in a single build. Sustainability is also a key focus, as AM reduces material waste and enables localized production, decreasing reliance on long supply chains.

As the AM technology has continuously evolved, it tends to lead to improved industrial production, new opportunities and the unlocking of new ways for innovation, productivity and feature-oriented developments in multiple sectors.

3D printing, also known as the fused filament fabrication (FFF) technique, consists of several different processes, each with their own method of creating objects layer by layer, as shown in Figure 3.1.

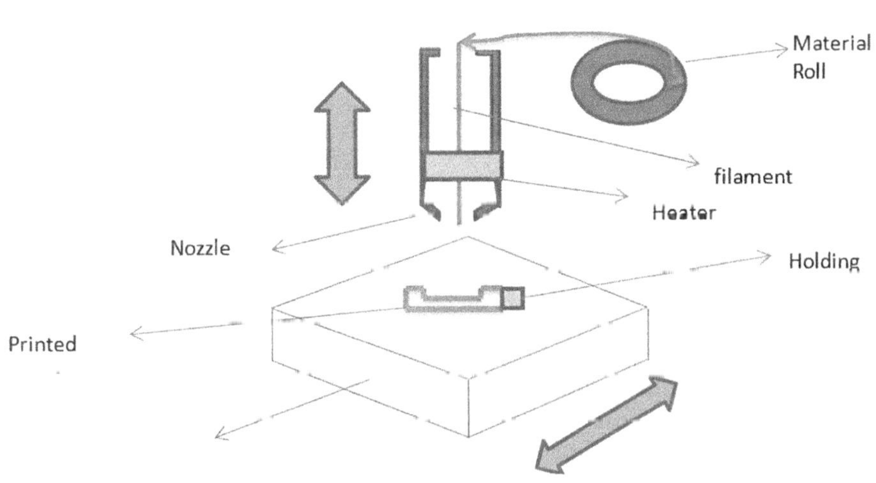

FIGURE 3.1 Principle of fused deposition modeling.

The characteristics of FDM are as follows:

- Process: In this process, thermoplastic materials (like polylactic acid (PLA), acrylonitrile butadiene styrene (ABS), or polyethylene terephthalate glycol (PETG)) are heated and extracted through the nozzle and form the desired material.
- Common materials: PLA, ABS, PETG, and thermoplastic polyurethane (TPU) (flexible)
- Applications: Prototyping, hobbyist projects, mechanical parts
- Benefits: Affordable, widely available, easy to use
- Limitations: Lower resolution, visible layer lines

3.2 IMPACTS OF VARIOUS PARAMETERS ON THE MECHANICAL PROPERTIES

The performance of the AM component depends on various key process parameters as their mechanical behavior and properties can be varied. These parameters include layer thickness, scan speed, laser power, build orientation, and cooling rate. They govern the micro-structure, anisotropy, presence of tiny pores, and residual stress in AM-fabricated components (Ramu et al., 2018). Visible layer lines are produced using ABS and PLA thermoplastics in a Stratasys F170 3D printer. Important parameters for 3D printing include the deposition layer thickness of the material, infill pattern, as well as angle, and orientation of the deposition material, and need to be carefully selected.

The prepared parts are then evaluated for their friction and wear behavior using a multi-tribo tester (TR-25) in a block-on-roller configuration. Tests are conducted in a dry condition with the same rotational speed against an EN 8 steel roller under low loads, with a fixed rolling time.

The results are systematically tabulated and graphed, accompanied by a detailed discussion to analyze the influence of the printing variables on the tribological and friction behavior of the 3D-printed ABS and PLA samples (Roy & Mukhopadhyay, 2021). Unlike traditional subtraction manufacturing, AM is found to provide flexibility in design, less wastage of material, and the ability to produce complex and intricate parts. However, the mechanical properties of AM components—such as tensile strength, ductility, hardness, and fatigue life—are highly sensitive to process parameters. So, we can say that the various process parameters directly impact on the reliability and performance of the AM part. The literature review explores how these parameters impact on the mechanical properties, using findings from power bed fusion (PBF), direct energy deposition (DED), and FFF techniques (Jamil et al., 2022a, 2022b).

An overview is provided of the mechanical properties and parameters that impact these properties in the case of the various AM techniques that are involved in building the parts. The parts are generally made up of metals and polymers. On the basis of that, a specific technique is employed. Each technique consists of various process parameters, and every parameter impacts the mechanical and behavioral characteristics of the fabricated part. In the case of metallic AM, various parameters are considered, such as spacing, hatching, electron beam power, and energy density; whereas, in the case of polymer AM, parameters such as build direction, nozzle and bed temperature, layer thickness, the orientation of the extracted material (Parnian et al., 2024), and infill density are critical. According to this, various properties like solidification time, material deposition rate, and energy input can be effected mechanically, chemically and at the micro-structural level as well. Many researchers report that, if these parameters are considered effectively, a part made by any technique like metal or polymer AM will have no defect, thereby increasing the mechanical performance of the final product. However, if these parameters are not considered effectively, it causes ineffective efficiency, impacting the mechanical strength, surface finish, and quality of the final product.

3.2.1 Key Process Parameters and Their Effects

3.2.1.1 Intensity of Laser and Energy Density for Metal AM

Studies indicate that higher energy density can reduce porosity by ensuring the complete melting of the powder, thereby enhancing tensile strength and ductility. For instance, research on Ti-6Al-4V components produced via selective laser melting (SLM) found that increasing energy density from 50 J/mm³ to 80 J/mm³ reduced porosity from 1.5% to 0.2%, improving ultimate tensile strength by approximately 15%. The excessive heat led to the keyhole defect, which impacted the fatigue strength of the material and its many other proprieties as well, which directly depend on heat. So, it is suggested that the energy should be at the optimal level to perform and get a higher output from the processes (Zhang et al., 2019).

3.2.1.2 Scanning Speed (Metal AM)

Scanning speed plays an important role in the consistency of the prepared part, and in the solidification time of manufacturing the part. A higher scanning speed can lead to a certain defect in the parts, like incomplete fusion due to the tensile and fatigue strength of the parts being impacted. However, if the scanning speed is slower than the overheating of the material, it will reduce the hardness of the material. Research on Inconel material revealed that a scanning speed of 600 mm/s yielded a fine microstructure with a hardness of 350 HV, while a scanning speed of 1200 mm/s resulted in a 10% drop in hardness due to insufficient melting.

3.2.1.3 Layer Thickness (Metal and Polymer AM)

Layer thickness is another important factor, which affects the tensile strength of the material. The thinner the layer, the lower the staircase effect will be, but this will result in a longer production time. This also impacts the build quality and inter-layer bonding between the material. Certain research on PLA shows that changing the thickness from 0.3 to 0.1 mm can increase the tensile strength up to 20%, at the same time doubling the printing time. Thicker layers, while being faster, often result in weaker bonding and lower mechanical strength, highlighting a trade-off between efficiency and performance.

3.2.1.4 Build Orientation (Polymer and Metal AM)

Build orientation is the direction in which the part to be prepared impacts significantly on the anisotropy of the material. In the case of AM, some research shows that a PLA-printed part with a 0° orientation (layers parallel to the tensile axis) exhibited 30% higher tensile strength than that printed at 90° (layers perpendicular) due to stronger inter-layer bonding. Similarly, using SLM with stainless steel 316L (Kumar et al., 2021), vertical builds showed superior ductility compared to horizontal builds. This can be attributed to the alignment of defects and grain boundaries. Anisotropy remains a key challenge in AM, as it complicates the prediction of mechanical behavior under varying loads. Figure 3.2 shows the build orientation.

3.2.1.5 Infill Density and Pattern (Polymer AM)

In FFF, infill density and pattern govern the internal structure and weight of the part. Studies on ABS components revealed that increasing infill density from 20% to 100% improved tensile strength from 25 MPa to 40 MPa, increasing the final manufacture time. Infill patterns, such as honeycomb or rectilinear patterns, also influence load distribution, with honeycomb patterns offering better compressive strength due to their uniform stress distribution.

3.2.1.6 Temperature Parameters (Polymer and Metal AM)

In FFF, nozzle and build plate temperatures affect material flow and adhesion. Higher nozzle temperatures (e.g. 220°C vs. 200°C for PLA) enhance filament bonding, increasing tensile strength by up to 15%, but excessive heat can degrade the polymer, reducing ductility. In metal AM, preheating the build platform reduces thermal

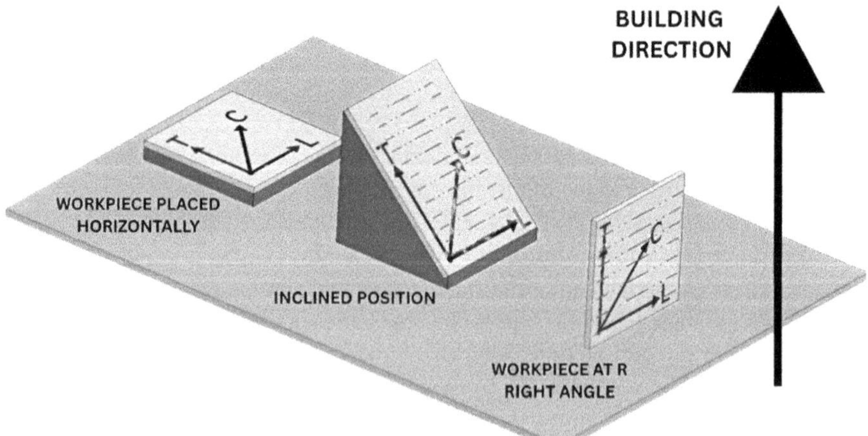

FIGURE 3.2 Build orientation for the material.

gradients, minimizing residual stresses and cracking. For example, preheating to 200°C in the SLM of Ti-6Al-4V improved fatigue life by 25% compared to non-preheated builds.

3.3 MECHANICAL PECULIARITIES AND DEFECTS

The mechanical peculiarities of AM components—such as anisotropy, porosity, and surface roughness—are direct outcomes of process parameter settings. Porosity is a common defect that can occur in the AM process. This happens due to the hot flue gas entrapped between the cavity of the product or due to a lack of fusion. The common reason for this is when an inappropriate layer thickness is selected, which causes the fatigue strength of the part to be reduced, as shown in Figure 3.3. Some researchers found that optimizing hatching can reduce porosity and increase fatigue strength to a large extent. For metal and non-metal products, like polymers, surface finish can play a critical role as it is directly dependent upon the layer thickness: as the thickness of the layer decreases, surface roughness is increased, but at the same time, the productivity is reduced due to there being more time to complete the part.

Anisotropy is another critical peculiarity, which will take place due to the repeated manufacture of a part layer by layer. In the case of metallic AM, defects and grain orientation depend on the build direction (configuration X, Y, Z), whereas in the case of polymer AM, parts become brittle as the bonding between the layers is weak, especially when the force is applied across the layer (Martínez-García et al., 2019). These peculiarities underscore the need for tailored parameter optimization to meet specific application requirements.

3.4 OPTIMIZATION APPROACHES AND CHALLENGES

To minimize the adverse effect of the process parameter, for instance, an artificial neural network (ANN) model could accurately predict the tensile strength of SLM-fabricated AlSi10Mg with an R^2 value of 0.95. However, concerning issues relate to finding the best geometry modeling for different metals and alloys, such as the high cost of computation data for simulations.

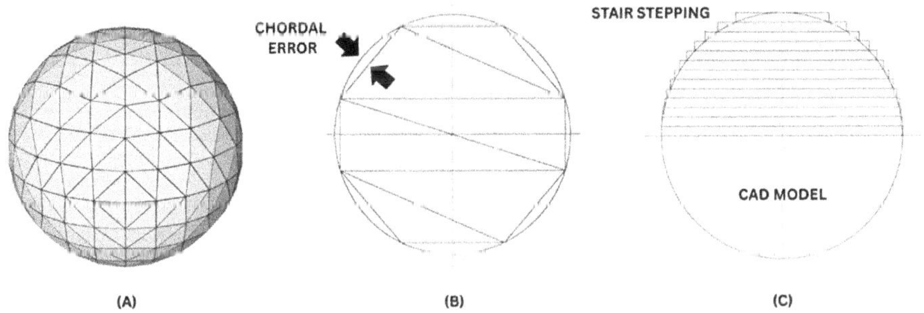

FIGURE 3.3 Defect produced during layer-by-layer manufacturing.

3.5 CONCLUSION AND RESEARCH GAPS

The literature reveals that the mechanical peculiarity of AM is highly dependent on the material, process parameters, and techniques through which the part is to be prepared. Plenty of research has already been done on the various relationships between the processes. Until now, limited studies have been carried out on the behavior of AM parameters, the multiple parameter interactions between them, industrial process planning, and optimization techniques or hybrid techniques. Future studies should be focused on the interdisciplinary techniques of optimization, and standardized operating procedures for developing and testing products so that the overall productivity and efficiency can be measured and improved effectively.

Layer thickness plays a vital role in finding the surface roughness and mechanical bonding of AM components. Studies indicate that a reduced layer thickness enhances the densification of materials and improves tensile strength. However, finer layers may increase manufacturing time and cost (Gibson et al., 2021). Layer thickness is considered to be the height of the layer during the pouring of the material layer by layer. In PBF techniques like SLM or electron beam melting (EBM), it refers to the thickness of the layer in powder form on the prepared bed before it is melted by the electron beam. In FFF, material extracted through the nozzle will provide the height of the thickness. Typical values range from 20–100 micrometers (μm) in metal AM and 0.1–0.4 millimeters (mm) in polymer AM, although these can vary depending on the machine, material, and desired outcome.

Layer thickness affects the resolution of the part directly, the strength of interlayer bonding, and the thermal history of the material, making it a critical determinant of mechanical behavior.

3.6 EFFECTS OF LAYER THICKNESS ON MECHANICAL PECULIARITIES

3.6.1 Interlayer Bonding and Tensile Strength

Inter-layering of the part is dependent on the thickness of the layer, which is considered to be part of the manufacturing process, as inter-layering is directly proportional to the adhesion between the two consecutive layers. The less the thickness, the greater the fusion between the layers. This also provides strength to the outer-fiber, thereby improving the fluxural strength. So, we can conclude that the layer plays an important role in the tensile strength and ductility of the product. For instance, a study on PLA components fabricated via FFF showed that reducing the thickness of the layer from 0.3 mm to 0.1 mm increased tensile strength by approximately 20%. This improvement stems from enhanced contact and diffusion between the layers, reducing weak interfaces that act as stress concentration points under load.

3.6.2 Surface Roughness and Fatigue Performance

A thinner layer of material will enhance the surface finish of the material and reduce defects like the "staircase effect". Discrete layering will improve the surface finish of the material. Figure.3.3. shows the staircase effect. Surface roughness is a critical factor in fatigue performance. In the SLM of Ti-6Al-4V, reducing the layer thickness from 60 μm to 30 μm improved the surface finish and increased fatigue life by up to 20%, as reported in studies (Xu et al., 2021). Conversely, thicker layers exacerbated surface irregularities, compromising wear resistance and fatigue strength, particularly in load-bearing applications.

3.6.3 Porosity and Defect Formation (Metal AM)

In metal AM, layer thickness affects both the energy input per layer and the cooling rate, influencing defect formation. The greater the thickness, the more energy will be required to melt the powder. Insufficient energy in the form of heat will cause lack-of-fusion porosity—voids between the layers that weaken the part. A study on AlSi10Mg fabricated via SLM found that increasing layer thickness from 30 μm to 60 μm increased the porosity from 0.5% to 1.2%, reducing tensile strength by 10%. Thinner layers, while reducing porosity, increase the number of layers, and thus the opportunities for minor misalignment or inconsistencies, necessitating the precise control of other parameters like laser power and scanning speed.

3.6.4 Anisotropy

In AM, the part is prepared in layers, which inherently introduces anisotropy. The layer thickness magnifies the effect of anisotropy. In the case of transverse loading, the thicker section produces a greater interface due to the section becoming weaker in that portion. In FFF, thicker layers (e.g. 0.3 mm) resulted in greater anisotropy in PLA parts, with tensile strength dropping significantly when loaded perpendicularly to the layer direction compared to thinner layers (e.g. 0.1 mm). In metal AM, anisotropy is more tied to grain orientation, but thicker layers can exacerbate defect alignment, further influencing directional properties.

3.6.5 Build Time and Efficiency

There is an inverse relationship between layer thickness and build time. As the layer thickness decreases, the build time to manufacture the product increases because more material is poured onto the surface in layers. As a result, more energy is required to do more work. For example, by halving the thickness from 0.2 to 0.1 mm, the time required to build the part is doubled for the same size product. For industrial uses, the productivity and efficiency increases, which leads to increasing the thickness of the layer up to the width that the design will allow.

3.6.6 Residual Stresses (Metal AM)

In the case of metal AM processes, a thicker layer of material is used, which cools and solidifies the molten filament in one go. This also increases the thermal gradient and volume of the material to resist the temperature, which increases the residual stress in the part. This residual stress leads to various defects like distortion, cracks, and delamination of parts for material like the Inconel 718 alloy. However, thinner layers do the opposite. Preheating the build platform can further alleviate this issue, but layer thickness remains a primary control factor (Sakib et al., 2024).

Trade-offs and optimization take place. The impact of layer thickness is categorically mentioned below:

- Thinner layers: Increase the tensile strength, build quality and surface finish, but at the same time, increase the time of production, and commutative error risk (e.g. misalignment over many layers).
- Thicker layers: Enhance productivity and reduce costs, but compromise mechanical properties, surface finish, and defect control.

Optimization often involves balancing these factors based on the component's application. For high-performance parts (e.g. aerospace components), thinner layers (20–50 μm in SLM, 0.1–0.2 mm in FFF) are preferred to maximize strength and durability. For prototypes or less critical parts, thicker layers (50–100 μm in SLM, 0.2–0.4 mm in FFF) may suffice to prioritize speed and cost (Guo et al., 2024).

Research has found that the integration of the design of the experiment and machine learning will provide the optimal solution for layer thickness. These studies indicate that the design of the experiment for the SLM of stainless steel 316L grade can be optimized by selecting and varying the laser scanning and power for a layer thickness of 40 μm, while providing the maximum tensile strength.

3.6.7 Scan Speed and Laser Power

Scan speed and laser power significantly influence the melting and solidification process in laser-based AM techniques, such as SLM and DMLS. If the speed is excessive, fusion will be incomplete, increasing porosity, while optimal laser power ensures better bonding between successive layers (DebRoy et al., 2018). Scanning speed and laser power profoundly influence the mechanical peculiarities of AM components by controlling energy delivery, defect formation, and micro-structural evolution. Higher laser power enhances strength and reduces porosity, but risks thermal issues, while faster scanning speeds boost efficiency, but can compromise fusion and fatigue life. Their interplay, encapsulated in energy density, dictates the trade-offs between productivity and performance. Optimal settings depend on the material, application, and other parameters, making their careful calibration essential for producing reliable, high-quality AM parts. Ongoing advancements in instantaneous monitoring and adaptive control promise to further refine their impact.

Practical Implications

For Polymer AM (FFF):

The best surface finish and optimal tensile strength is achieved at 0° surface horizontal if the angle is modified from 0°–90° with the surface vertical, its provided slender parts generally being taller in section, but weaker under transverse loads

(e.g. columns) if the angle is shifted to 45° with the surface compromised for complex geometries (e.g. brackets).
For Metal AM (SLM):
With an angle of 0° to the surface (horizontal): Preferred for high tensile strength and minimal supports (e.g. beams)
With an angle of 90° to the surface (vertical): Enhances ductility and fatigue life for specific loads (e.g. shafts).
With an angle of 45° to the surface: Reduces anisotropy and defect impact (e.g. aerospace fittings) (Pekok et al., 2021).

3.6.8 Rate of Cooling and Post-processing Method

The cooling rate plays a vital role in pre- and post-processing techniques. The cooling rate is high, it may lead to a brittle structure. If it is in a controlled condition, it may lead to ductility in the structure. Heat treatment processes like annealing and constant temperature pressing such as hot isostatic pressing (HIP) help mitigate residual stress, which is present in the material, thereby enhancing the mechanical properties (Liu et al., 2020). Solidification time influences various important parameters, micro-structures, and mechanical properties like the hardness, ductility, toughness, and impact strength of the AM part.

- Micro-structure: Higher solidification time (e.g. from high scanning speeds) refines the grain size of parts, enhancing the parts' mechanical properties. For the SLM of Inconel 718, a scanning speed of 1200 mm/s reduced the grain size, increasing hardness by 10% compared to 600 mm/s. Slower solidification time (e.g. low speeds or pre-heated platforms) leads to coarse grain sizes, enhancing ductility, but reducing hardness.
- Defects and stresses: Faster cooling leads to certain defects like porosity, cracks, and an increase in the gradient. However, slower cooling, as with a preheated bed at 200°C, reduces stress by 25% and improves fatigue life.
- Mechanical properties: In the case of the FDM of polymers, in particular, rapid cooling reduces the strength between the layers, thereby lowering the tensile strength (e.g. with PLA at 20% less strength), while controlled cooling increases the adhesion between the layers.

3.7 MECHANICAL CHARACTERIZATION

Mechanical characterization is essential for understanding material behavior and for evaluating performance and reliability under different loading and unloading conditions. Tensile testing is one of the fundamental techniques used to determine a material's resistance to uniaxial loading. The tensile test is one of the highly performed tests on various materials to check their properties, such as Young's modulus of elasticity, yield point, breaking point, and elongation against the load and ductility of the material, and its ability to reform in shape under certain load conditions. It therefore plays an important role in shaping and reshaping the reinforced material (Alomarah et al., 2022). There are numerous methods of checking all these material tensile test parameters . It is best to consider all of them due to the simplicity of the test.

As mentioned above, the tensile test is considered the best method to determine the various mechanical properties of material using specimens. On unidirectional loading, this means two opposites pull or tension takes place until failure occurs. This test is generally used to find mechanical properties like the strength of the material, yield points, and elongation in terms of the stress strain analysis method. Two standard methods are employed on polymers to check their properties, namely the ASTM D638 method, which is used for polymers and composite materials, and the ASTM E8 method, which is used for metals and alloys. The biggest benefits of these methods are that they will provide the standard operation to determine the material's various properties. They are also suitable to calculate the orientation and behavior of materials under different load conditions (Akbari et al., 2022).

Flexural testing is one of the tests used to find the bending strength of material without failure. It is typically used in two types of configurations: namely three points, and four points. Flexural testing provides the strength of the outer fiber under stress conditions that generally occur in polymer and polymer-based reinforced material, so these properties become significant for polymer material. It also provides the flexural modulus, which provides information about the material that can bear a load under resistance (stiffness) during bending. It plays an important role in AM as it reveals the structural strength of the product and avoids the presence of any type of void on the surface. It also reveals the strength of the outer layer of the additive parts under bending, and provides information about the orientation of the product and its structural strength.

Hardness testing is a technique used to represent the strength of material against scratch indentation, and penetration against the load. For metallic materials, standardized methods such as Vickers, Rockwell, and Brinell tests are commonly adopted, while Shore D hardness testing is generally applied for polymers and composites. In the context of AM, hardness measurements serve as an important tool to assess surface integrity, micro-structural variations, and the influence of processing parameters. Unlike bulk mechanical tests, hardness testing can provide localized information, making it particularly useful for characterizing the anisotropy and heterogeneity often observed in AM parts (Bhardwaj et al., 2022). Variations in hardness may arise from differences in cooling rates, layer-wise thermal history, residual stresses, and some defects due to incomplete fusion for porosity, for example. Furthermore, hardness mapping across the build direction is an effective way to detect layer-dependent property variations. Various heat treatment processes, such as HIP, or surface finishing can also be evaluated through hardness testing, as improvements in densification or micro-structural refinement are often reflected in higher and more uniform hardness values. Although hardness is only an indirect measure of mechanical performance, it is strongly correlated with strength and wear resistance, making it a valuable indicator for assessing the reliability and service life of AM components.

3.8 MICRO-STRUCTURAL AND MORPHOLOGICAL CHARACTERIZATION

3.8.1 Optical Microscopy

Optical microscopy (OM) is a technique for examining the micro-structure of materials using visible light and magnifying lenses. In additive manufacturing, OM is employed to evaluate surface morphology, grain structure, porosity distribution, and inter-layer fusion. In AM, a specimen is prepared by polishing and rubbing it. The grain structure and phase distribution are influenced by rapid cooling rates during fabrication. Optical microscopy is particularly valuable for quality assessment as it provides a quick and cost-effective method to detect process-induced irregularities such as lack of fusion, delamination, or large pores. However, its resolution is limited compared to electron microscopy, restricting its ability to capture nanoscale features. Despite this limitation, OM remains an indispensable tool for routine evaluation and preliminary micro-structural analysis in AM research.

3.8.2 Scanning Electron Microscopy

Scanning electron microscopy (SEM) provides a high-resolution image of the material surface using a focus electronics beam. In additive manufacturing, SEM is widely used to analyze surface morphology, inter-layer bonding, micro cracks, unmelted powders, and porosity. Back-scattered electron imaging can highlight compositional contrast, while energy-dispersed X-ray spectroscopy (EDS), when coupled with SEM, enables the elemental analysis of regions of interest. SEM is particularly valuable for understanding the micro-structural heterogeneity caused by different process parameters, as well as for characterizing the fracture surfaces of tensile, flexural, or impact-tested AM specimens. The main advantages of SEM are its high magnification, excellent depth of field, and ability to reveal fine-scale defects, although it requires conductive coatings for polymers and has relatively high preparation costs.

3.8.3 Transmission Electron Microscopy

Transmission electron microscopy (TEM) allows imaging at the nanometer scale by transmitting a beam of electrons through an ultrathin sample. In AM research, TEM provides insights into nanoscale micro-structural features such as dislocations, nano-precipitates, fine grains, and amorphous regions formed due to rapid solidification. Selected area electron diffraction (SAED) in TEM is particularly useful for identifying crystallographic orientations and phases present within AM alloys (Gökdağ & Demir, 2022). While TEM offers unmatched resolution for analyzing structural evolution at the atomic level, it is highly time-consuming, expensive, and requires complex sample preparation, which limits its use mainly to advanced research studies rather than routine quality control.

3.8.4 X-ray Diffraction

The technique of X-ray diffraction (XRD) is used to find the bonding between the material, and checking the internal structure, residual stress, and composition of the material. In case of AM components, XRD helps to detect the type of phase formation during the rapid cooling of the material in the pre- and post-processing of prepared parts. This technique provides a fingerprint of the material's crystal structure, making it especially valuable in studying phase stability in alloys and composites used in AM (Gökdağ & Demir, 2022). XRD is a bulk technique and does not provide spatial information on micro-structural defects, which is why it is often combined with microscopy methods for comprehensive characterization.

3.8.5 X-ray Computed Tomography

X-ray computed tomography (XCT/CT-scanning) uses multiple X-ray projections to generate the 3D image of an internal structure. In the case of AM, this method is adopted to check various defects like porosity, cracks, and lack of fusion without the need for destructive sectioning. Its use of volumetric analysis makes it ideal for assessing quality in aerospace and biomedical applications, where reliability is critical. XCT can also be used to validate process parameters by correlating defect distribution with printing conditions. The primary advantage of XCT is its ability to capture hidden defects, but limitations include its high cost, long scanning times, and lower resolution compared to SEM or TEM for nanoscale features.

The finishing and polishing of the finished part prepared by AM will diminish the defect that is generated through the process of making parts, like improper surface finish and roughness. Thereby it increases the productivity and mechanical performance of the process (Bhuvanesh Kumar et al., 2023).

- Stress relief: Heat treatment reduces residual stress in metal parts prepared by AM (e.g. annealing Ti-6Al-4V lowered stresses by 30%), improving fatigue life and dimensional stability.
- Surface quality: Machining or polishing reduces surface roughness (e.g. from Ra 10 μm to 2 μm in SLM stainless steel), boosting fatigue resistance by up to 25%.
- Micro-structure refinement: Post-heat treatment of AlSi10Mg in SLM homogenized the micro-structure, increasing ductility by 15%, while maintaining strength.
- Defect mitigation: Hot isotactic pressing closes internal porosity in metal AM, raising tensile strength (e.g. by 10% in Inconel 718).

3.9 COMPARATIVE ANALYSIS OF AM TECHNIQUES

Different AM techniques exhibit varied mechanical properties due to differences in material deposition and energy input. PBF processes such as SLM provide high strength and density, while FDM components often suffer from weak interlayer adhesion. Binder jetting and DED offer unique advantages, but require careful parameter control to achieve the desired mechanical characteristics (Guo & Leu, 2021).

3.10 SUMMARY AND FUTURE SCOPE

The additive manufacturing process is widely spread all over the world due to its ease of operation and various advantages over the traditional method of manufacturing. In the case of AM, various parameters that are selected for the process of fabrication play a major role in deciding on their efficiency, accuracy, and precision. In a similar fashion, various properties, like the mechanical, rheological, morphological, and behavioral characteristics, also depend on the process parameters that are chosen for the operation. Future research should be focused on the real-time monitoring of these parameters and adaptive control in the operation. Machine learning and simulation-driven optimization techniques, artificial neural networks, computational neuro-networking, and various optimizing tools and techniques enhance the ability to predict, and thereby enhance, the capability of predictive maintenance and the requirement of computer-aided process planning, and predictive modeling for AM process control.

REFERENCES

Akbari, P., Zamani, M., & Mostafaei, A. (2022). MechProNet: Machine learning prediction of mechanical properties in metal additive manufacturing. *arXiv preprint*. https://doi.org/10.1016/j.addma.2024.104320

Alomarah, A., Abbas, A. T., Faisal, B., Peng, Z., & Ruan, D. (2024). The effects of manufacturing techniques on the mechanical performance of an auxetic structure manufactured by fused filament fabrication and multijet fusion processes. *Advanced Engineering Materials*, 26(7), 2302033. https://doi.org/10.1002/adem.202302033

Bhardwaj, N., Henein, H., & Wolfe, T. (2022). Mechanical properties of thermoplastic polymers in fused filament fabrication (FFF). *The Canadian Journal of Chemical Engineering*, 100(11), 3174–3183. https://doi.org/10.1002/cjce.24562

Bhuvanesh Kumar, M., Sathiya, P., & Senthil, S. M. (2023). A critical review of wire arc additive manufacturing of nickel-based alloys: Principles, process parameters, microstructure, mechanical properties, heat treatment effects, and defects. *Journal of the Brazilian Society of Mechanical Sciences and Engineering*, 45, 164. https://doi.org/10.1007/s40430-023-04077-1

DebRoy, T., Wei, H. L., Zuback, J. S., Mukherjee, T., Elmer, J. W., Milewski, J. O., ... & Zhang, W. (2018). Additive manufacturing of metallic components process, structure and properties. Progress in Materials Science, 92, 112–224. https://doi.org/10.1016/j.pmatsci.2017.10.001

Gibson, I., Rosen, D., Stucker, B., & Khorasani, M. (2021). *Additive Manufacturing Technologies* (pp. 160–186). Cham, Switzerland: Springer. https://doi.org/10.1007/978-3-030-56127-7

Gökdağ, İ., & Demir, A. G. (2022). Effects of process parameters on strengthening mechanisms of additively manufactured AlSi10Mg. *Materials Testing*, 64(5), 487–494. https://doi.org/10.1515/mt-2022-0449

Guo, L., Li, L., Li, Y., Duan, X., Liu, Y., Gao, R., & Zhao, Y. (2021). Characterization of antimicrobial resistance and virulence genes of

Pseudomonas aeruginosa isolated from mink in China, 2011–2020. *Microbial Pathogenesis*, 162, 105323. https://doi.org/10.1016/j.micpath.2021.105323

Guo, X., Yu, Y., Allon, G., Wang, M., & Zhang, Z. (2024). Ririshun logistics: Home appliance delivery data for the 2021 manufacturing & service operations management data-driven research challenge. *Manufacturing & Service Operations Management*, *26*(4), 1358–1371. https://doi.org/10.1287/msom.2021.0994

Jamil, I. R., Mustaquim, A. M., Islam, M., & Hasan, M. N. (2022a). Understanding mechanical characteristics of FeNiCrCoCu HEA in nanoscale laser powder bed fusion via molecular dynamics. arXiv preprint. https://arxiv.org/abs/2207.08091

Jamil, I. R., Mustaquim, A. M., Islam, M., Thakur, M. S. H., & Hasan, M. N. (2022b). Effect of powder bed fusion process parameters on microstructural and mechanical properties of FeCrNi MEA: An atomistic study. arXiv preprint. https://doi.org/10.1007/s12540-022-01263-z

Kumar, R., Kumar, J., & Prakash, O. (2021). A critical review on effect of process parameters on mechanical and microstructural properties of powder-bed fusion additive manufacturing of SS316L. *Materials*, 14(21), 6527. https://doi.org/10.3390/ma14216527

Fayazfar, H., Salarian, M., Rogalsky, A., Sarker, D., Russo, P., Paserin, V., & Toyserkani, E. (2018). A critical review of powder-based additive manufacturing of ferrous alloys: Process parameters, microstructure and mechanical properties, *Materials & Design*, 144, 98–128, https://doi.org/10.1016/j.matdes.2018.02.018.

Liu, C., Mai, Z., Yan, D., Jiang, M., Dai, Y., Wang, P., ... & Lao, C. (2020). Effect of hot isostatic pressing on microstructures and mechanical properties of Ti6Al4V fabricated by electron beam melting. *Metals*, *10*(5), 593. https://doi.org/10.3390/met10050593

Martínez-García, A., Sandoval-Pérez, I., Ibáñez-García, A., Pernías-Peco, K., Varela-Gandía, F. J., & Galvañ-Gisbert, J. (2019). Influence of process parameters of different additive manufacturing techniques on mechanical properties and safety of customised toys. *Procedia Manufacturing*, 41, 106–113. https://doi.org/10.1016/j.promfg.2019.07.035

Parnian, P., Russo, P., Squillace, A., & D'Amore, A. (2024). The process-structure–property relationship of 3D printed G-Polymer using fused filament fabrication technique. *Polymer Engineering & Science*, 64(7), 3073–3087. https://doi.org/10.1002/pen.26748

Pekok, M. A., Setchi, R., Ryan, M., & Liu, Y. (2021). Effect of process parameters on the microstructure and mechanical properties of AA2024 fabricated using selective laser melting. *The International Journal of Advanced Manufacturing Technology*, 112(1), 175–192. https://doi.org/10.1007/s00170-020-06346-y

Ramu, M., Mitilesh, R. N., & Singamneni, S. (2018). Influence of process parameters on the mechanical behaviour and processing time of 3D printing. *International Journal of Modern Manufacturing Technologies*, 10(1), 69–75.

Roy, R., & Mukhopadhyay, A. (2021). Tribological studies of 3D printed ABS and PLA plastic parts. *Materials Today: Proceedings*, *41*, 856–862. https://doi.org/10.1016/j.matpr.2020.09.235

Sakib, T., Enemuoh, E. U., Houle, T., & Bauer, L. (2024). Impact of process parameters on mechanical properties and surface characteristics in hybrid metal additive manufacturing of maraging steel. *The International Journal of Advanced Manufacturing Technology*, 135, 4559–4575. https://doi.org/10.1007/s00170-024-14704-3

Xu, J., Li, H., & Gao, G. (2021). The effect of 3D printing process parameters on the mechanical properties of PLA parts. *Journal of Physics: Conference Series*, 2133(1), 012026. https://doi.org/10.1088/1742-6596/2133/1/012026

Zhang, Y., Gao, M., & Zeng, X. (2019). Effect of process parameters on mechanical properties of wire and arc additive-manufactured AlCu6Mn. *JOM*, 71(3), 886–892. https://doi.org/10.1007/s11837-018-3210-8

4 Repercussion of Process Parameters on the Mechanical Peculiarity of Additive Manufactured Components

Sulaim Rizvi, Shahazad Ali, Anant Prakash Agrawal, and Kashif Irshad

4.1 INTRODUCTION

Additive manufacturing (AM) has revolutionized the contemporary production process and provided a possibility that the traditional manufacturing process has failed to offer (Attaran, 2017; Gibson et al., 2020). In contrast to traditional manufacturing methods, which are subtractive, AM is additive in nature, and uses the layer-by-layer addition of material to develop components based on digital 3D models. This allows complex shapes to be created with minimal material wastage (Ngo et al., 2018; Tofail et al., 2018). One of the AM processes, fused filament fabrication (FFF), has also gained significant popularity because it is economical, available, and can handle a range of thermoplastics (Dizon et al., 2018; Popescu et al., 2018).

The thermoplastic material that has been popular in FFF is acrylonitrile butadiene styrene (ABS) as it has good mechanical properties, thermal stability, and post-processing capabilities (Wang et al., 2017; Wickramasinghe et al., 2020). Nevertheless, parts produced by FFF generally exhibit anisotropic behavior and have low mechanical properties compared to those produced by conventional technologies (Rankouhi et al., 2016; Torrado & Roberson, 2016). This is due to the layer-by-layer deposition method, which forms natural interfaces between the layers and leaves possible voids inside the structure (Goh et al., 2020; Rodríguez et al., 2001).

The mechanical performance of FFF-printed components depends on several process parameters, such as nozzle temperature, layer thickness, print speed, and cooling rates (Algarni, 2021; Chacón et al., 2017; Liu et al., 2017). Nonetheless, there are three paramount measurements that greatly influence mechanical behavior, namely,

DOI: 10.1201/9781003643531-4

print orientation, infill density, and filling pattern (Carneiro et al., 2015; Sood et al., 2010; Tanveer et al., 2021):

- Print orientation is the track of deposition of the layer compared to loading direction, which influences interlayer bonding and anisotropic behavior.
- Pattern density suggests the ratio of material occupied within the part; this has a major effect on the weight of the components and their mechanical characteristics.
- Geometry of patterns determines the internal geometry of the structure, which influences both the pattern stress distribution and the pattern energy absorption ability.

Although the above parameters have been discussed individually in previous studies (Akhoundi & Behravesh, 2019; Cantrell et al., 2017; Fernandes et al., 2018; Kumar et al., 2019; Kuznetsov et al., 2018; Mishra et al., 2021), more comprehensive studies are needed to understand the combined impact of these parameters (Lubombo & Huneault, 2018; Rajpurohit & Dave, 2018). Furthermore, tensile properties are the center of attention of most studies, and little has been mentioned about flexural and impact behavior, which are also important in engineering usage (Alafaghani et al., 2017; Caminero et al., 2019).

The purpose of this work is to determine the influence of layer thickness, print density, and printing pattern on the tensile, flexural, and impact strength of ABS-CF components that have been developed through FFF. This study can be valuable to manufacturers who want to enhance the mechanical performance of ABS-CF specimens developed through FFF.

4.2 MATERIALS AND METHODS

4.2.1 Materials

For the development of the specimen, commercial ABS+CF filament with 1.75ϕ was used. Table 4.1 lists the material attributes that the manufacturer specified. To reduce

TABLE 4.1
Properties of ABS+CF Filament

Material characteristics	Measured values
Density (mass per unit volume)	1.10 g/cm³ (1100 kg/m³)
Ultimate tensile strength	48.0 MPa (6960 psi)
Maximum elongation at break	3.0%
Elastic modulus	5200 MPa (754,000 psi)
Flexural strength (bending stress)	78 MPa (11,310 psi)
Flexural modulus (bending stiffness)	5280 MPa (765,600 psi)
Glass transition temperature	105 (221°F)

moisture absorption, the filament was stored in a vacuum-sealed holder containing desiccant before printing.

4.3 DESIGN AND FABRICATION

American Society for Testing and Materials (ASTM) guidelines were followed in the design of the test specimens. Computer-aided design (CAD) was used to produce the specimens, which were then exported as STL files. The "Smart One Plus" fused deposition modeling (FDM) 3D printer was used to develop each specimen. The variable factors examined in this work are described in Table 4.2. The 3D printing was done with a nozzle diameter of 0.40 mm. The initial layer height was 0.27 mm. The line width was kept at 0.35 mm. The extrusion and build plate temperatures were 220°C and 100°C, respectively. The printing speed was maintained at 65 mm per second and the rapid movement speed was 80 mm per second so as to be able to deposit the layers efficiently and allow the print head to move smoothly (Agrawal et al., 2023; Raja et al., 2022; Srivastava et al., 2025).

The L9 array was used to get a better insight with a smaller number of experiments. The L9 array design used for this study is presented in Table 4.3, with three specimens per combination for each test type (tensile, flexural, and impact), resulting in a total of 27 specimens for each test (nine combinations with three replicates).

Compared to a full factorial design, this L9 orthogonal array design drastically reduced the number of experimental runs needed, while enabling an effective analysis

TABLE 4.2
Variable Parameters and Their Levels

Symbol	Parameter	Levels
A	Layer thickness	0.1–0.2–0.3 millimeter
B	Pattern density	75%–85% 95%
C	Printing pattern	Honeycomb–triangular–brick

TABLE 4.3
Taguchi L9 Orthogonal Array

Trial	Pattern	Infill density	Layer thickness
1	Honeycomb	75%	0.1 mm
2	Honeycomb	85%	0.2 mm
3	Honeycomb	95%	0.3 mm
4	Triangular	75%	0.2 mm
5	Triangular	85%	0.3 mm
6	Triangular	95%	0.1 mm
7	Brick	75%	0.3 mm
8	Brick	85%	0.1 mm
9	Brick	95%	0.2 mm

of the three parameters at three levels each. The primary effects could be statistically analyzed without confounding interactions, since the experimental matrix is balanced so that each level of each factor appears the same number of times.

4.4 MECHANICAL TESTING

4.4.1 Tensile Testing

The specimens were developed according to ASTM-D638 Type-I for the tensile strength assessment. Figure 4.1 shows the specimen, along with the dimensions in accordance with the ASTM Standard. A computerized Universal Testing Machine (UTM) was used at Sharda University in Uttar Pradesh, India, to assess the tensile strength. The test was performed using the AMT-10 model from Innotech Engineering Devices Pvt. Ltd., which has a load-bearing capacity of 3 tons, and a temperature-controlled chamber for maintaining temperatures between -70°C and 180°C, with a precision of ±1°C. Specimens were mounted in the testing machine with a 115 mm grip separation. The test was conducted at a speed of 2 mm per minute, and data for ultimate tensile strength (UTS), yield strength (YTS), and percentage elongation at the point of break were recorded.

4.4.2 Flexural Testing

For flexural strength of the developed specimen was estimated using the three-point bending tests according to the ASTM D790 Standard on the UTM. The specimen dimensions are shown in Figure 4.2. During testing, the crosshead moved steadily at a speed of 2.0 mm per minute, continuing until the outer surface strain reached 5% or the sample broke, whichever happened first. Flexural strength and flexural modulus were calculated using the load-deflection data sheet.

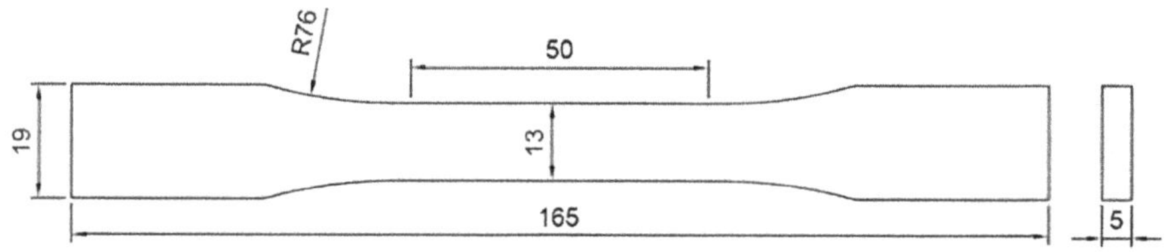

FIGURE 4.1 Tensile test specimen.

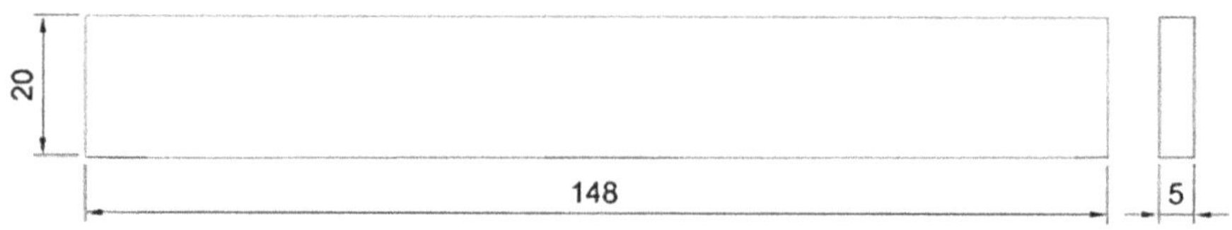

FIGURE 4.2 2D drawing of flexural test specimen.

4.4.3 Impact Testing

According to the ASTM D256 guidelines, specimens were made for the impact strength test. The specimens were designed on a CAD model, along with the dimensions shown in Figure 4.3. An impact testing machine with a maximum rating of 50 J was used for the impact tests. The specimens were positioned vertically in the fixture throughout testing, with the notch aligned to face the impact side of the pendulum. Impact strength was determined by dividing the energy absorbed before breaking (in kilojoules) by the specimen's cross-section area (in square meters) after the pendulum struck it.

4.5 RESULT AND DISCUSSION

4.5.1 Tensile Properties

Table 4.4 presents the tensile test data, performed on all nine samples. For every sample configuration, the table shows the elongation at break (ε_b) and UTS values, as well as the corresponding standard deviations from three replicate tests.

The ABS+CF specimens exhibited a strong dependence of tensile performance on printing process parameters, with UTS values spanning 31.6 MPa to 42.10 MPa,

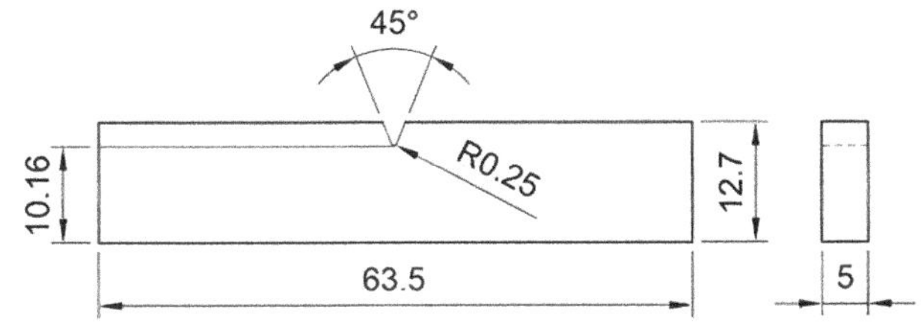

FIGURE 4.3 Impact testing specimen.

TABLE 4.4
Tensile Test Data

Sample ID	Pattern	Print density	Layer height	Ultimate tensile strength (MPa)	Elongation at break (%)
S1	Honeycomb	75%	0.1 mm	36.2 ± 1.5	2.5 ± 0.2
S2	Honeycomb	85%	0.2 mm	39.6 ± 1.6	2.8 ± 0.2
S3	Honeycomb	95%	0.3 mm	42.1 ± 1.7	3.0 ± 0.3
S4	Triangular	75%	0.2 mm	33.8 ± 1.5	2.3 ± 0.2
S5	Triangular	85%	0.3 mm	37.0 ± 1.6	2.6 ± 0.2
S6	Triangular	95%	0.1 mm	39.4 ± 1.7	2.7 ± 0.2
S7	Brick	75%	0.3 mm	31.6 ± 1.4	2.1 ± 0.2
S8	Brick	85%	0.1 mm	33.2 ± 1.5	2.3 ± 0.2
S9	Brick	95%	0.2 mm	35.9 ± 1.6	2.5 ± 0.2

and elongation at break point ranging from 2.1 to 3.0%. Increased infill densities and increased layer thickness resulted in an improved mechanical performance that was indicative of enhanced material continuity, interlayer bonding, and enhanced load transfer. The ABS matrix was reinforced with carbon fiber that offered a starting point of moderate ductility and high strength in all the configurations, although the geometry and processing parameters affected the performance envelope.

Out of the three infill patterns that were tested (honeycomb, triangular, and brick), the honeycomb design always performed better than the other two. The tensile strengths were the highest (36.2 MPa at 75% infill/0.1 mm layer to 42.1 MPa at 95% infill/0.3 mm layer) and had the highest elongations (up to 3.0%). By contrast, the triangular pattern delivered intermediate values (33.8 MPa to 39.4 MPa; 2.3 to 2.7% elongation), while the brick pattern showed the lowest strengths (31.6 MPa to 35.9 MPa) and ductility (2.1 to 2.5%). These differences underline how the internal stress distribution and number of load-bearing struts in each pattern control failure initiation and propagation.

Figure 4.4 and Figure 4.5 show that, with increasing infill density from 75 to 95%, the samples produced substantial gains in UTS and modest improvements in elongation across all patterns. For example, within the honeycomb series, boosting density from 75 to 85% raised the UTS by 3.4 MPa, and further increasing it to 95% added another 2.5 MPa. Denser prints harbor fewer voids and more continuous fiber matrix regions, enhancing stiffness and strength, although the incremental benefit begins to taper off above roughly 85%, suggesting diminishing returns in material use versus mechanical gain.

A similar effect is shown by layer thickness. The UTS climbed steadily as the height of the layer increased from 0.1 to 0.3 mm, again most prominently in the honeycomb samples (36.2 MPa → 39.6 MPa → 42.1 MPa). Thicker layers were found to reduce the number of interfacial boundaries and thermal gradients between deposited layers, fostering stronger fusion and fewer interlaminar defects. However, beyond 0.2 mm, the marginal strength improvements diminished, and excessive layer thickness could compromise surface fidelity and dimensional accuracy. Thus, a layer

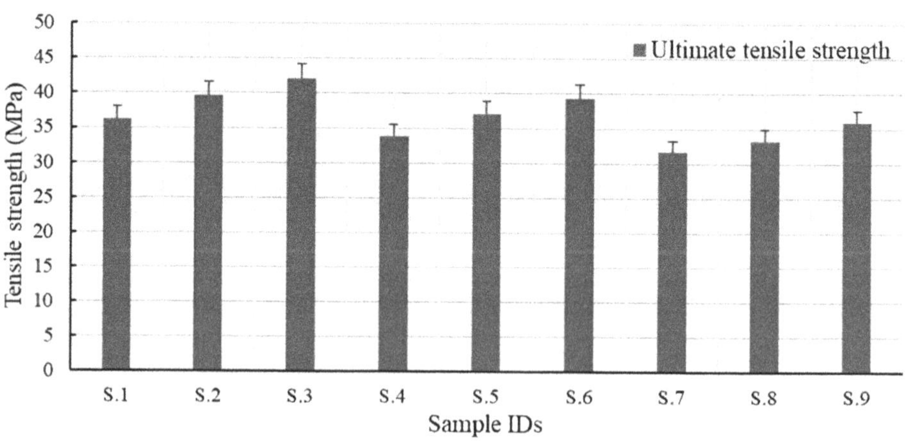

FIGURE 4.4 Effect of printing process parameter on tensile strength.

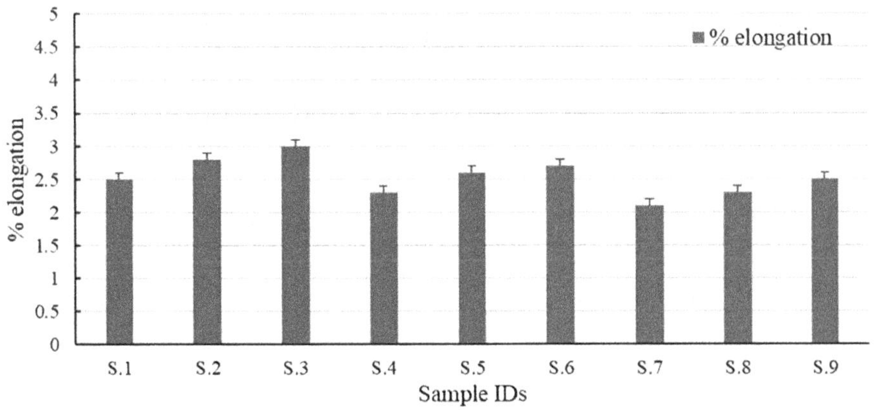

FIGURE 4.5 Effect of printing process parameter on percentage elongation.

height of approximately 0.2 mm is considered the best trade-off between mechanical properties and print quality for structural ABS+CF components.

4.5.2 Flexural Properties

All nine samples were tested by flexural tests. The findings are given in Table 4.5. The table represents flexural strength, flexural modulus, along with standard deviation out of three replicate tests. The flexural performance of the ABS+CF specimens was observed to be extremely susceptible to the printing structure and printing parameters. The flexural strength was recorded to be between 27.0 and 72.0 MPa. The general trend showed that the pattern density and layer thickness had great effects on the mechanical response to the bending loads. In addition, the nature of the internal pattern was also a major factor that influenced the distribution and resistance of stresses to flexural deformation. The ABS with added carbon fiber increased the stiffness of the material, and with optimized print settings, the flexural strength was high in a number of configurations.

Data from Table 4.5 and Figure 4.6 shows that a brick pattern enables impressive flexural strength in comparison to the honeycomb and brick patterns in the developed specimens. In the triangular samples, the measured values of flexural strength ranged from 59.0 MPa at 75% infill to 72.0 MPa at 85% infill and 0.3 mm layer thickness. Geometric rigidity and straight-line alignment of its struts during the triangular form was probably the factor that made it better resistant to bending loads. By contrast, honeycomb samples had a poorer flexural strength (27.0 to 38.0 MPa), presumably because the internal structure was rounded and so more deformable, which is good with tensile load, but not with flexural stress. The brick pattern showed superior results to the honeycomb pattern, but had a poorer performance than the triangular one, perhaps because of inefficient stress paths and reduced rigidity.

The influence of infill density was mostly positive in all the patterns where flexural strength was greater as the densities were higher. It was observed that, in the

TABLE 4.5
Flexural Test Results for All Experimental Samples

Sample ID	Pattern	Infill density	Layer thickness	Flexural strength (MPa)
S1	Honeycomb	75%	0.1 mm	27.0 ± 1.5
S2	Honeycomb	85%	0.2 mm	32.3 ± 1.6
S3	Honeycomb	95%	0.3 mm	38.0 ± 1.7
S4	Triangular	75%	0.2 mm	59.0 ± 2.0
S5	Triangular	85%	0.3 mm	72.0 ± 2.5
S6	Triangular	95%	0.1 mm	63.5 ± 2.2
S7	Brick	75%	0.3 mm	34.0 ± 1.6
S8	Brick	85%	0.1 mm	32.0 ± 1.5
S9	Brick	95%	0.2 mm	37.0 ± 1.8

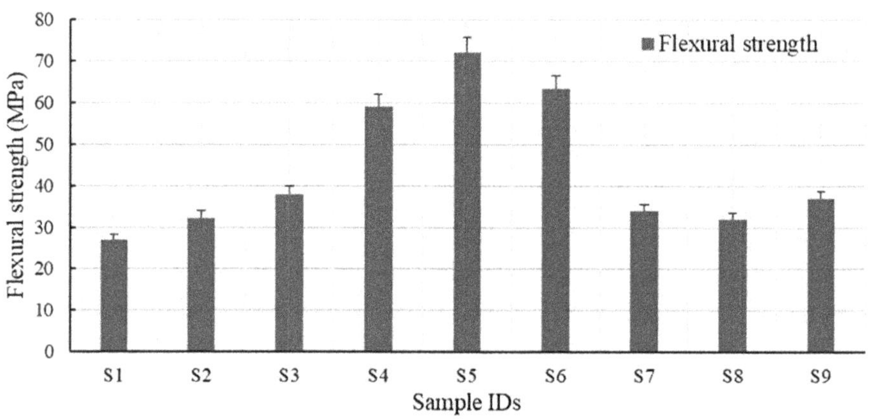

FIGURE 4.6 Influence of printing process parameter on flexural strength.

honeycomb samples, as the infill increased to 95%, the strength increased steadily to 38.0 MPa. The same trend was observed with the triangular- and brick-shaped infills. The triangular shape increased the most significantly between 59.0 MPa at 75% and 72.0 MPa at 85%, indicating a 13 MPa increase in resistance to the bending moments when using the denser structure. The triangular printing pattern, with 95% print density (S6), was a little weaker (63.5 MPa) than at 85% (72.0 MPa), indicating that over-compaction perhaps contributed to either brittle or inadhesive layers. This shows that there is an optimal level of density beyond which performance can be either flat or decreasing.

The flexural behavior was also influenced by layer thickness, whereby the greater the thickness of the layer, the greater the strength, particularly in the honeycomb and brick designs. In the case of the honeycomb samples, a distinct increasing trend was found with an increase in layer thickness. It went up to 0.10 mm (27.0

MPa) and 0.3 mm (38.0 MPa). The flexural strength was maximized at 0.3 mm in the brick samples (with a flexural strength of 34.0 MPa), which is probably due to the increased bonding between the layers, minimizing the effect of thermal contraction, and increased the structural stiffness of the samples during the bending process. However, in the triangular samples, the relation was not so linear. The maximum strength peaked at 0.3 mm (72.0 MPa), and the 0.1 mm sample (63.5 MPa) was again more effective than the 0.2 mm (59.0 MPa) sample. The specific interaction with layer thickness could have been more significant than a general pattern.

4.5.3 Impact Properties

Table 4.6 presents the results of the impact. Density, infill pattern, and layer thickness also influenced the impact behavior of the ABS+CF specimen. An optimized change in layer thickness and infill density normally led to improved impact resistance in all configurations. Carbon fiber reinforcement was instrumental in the increase in the energy absorption capacity, particularly in the presence of optimized print parameters, which encourage good interlayer bonding and internal structural integrity. The trends were both reasonable and promising to applications of toughness and energy dissipation, although the improvements were only in the tensile or flexural properties' magnitude.

Out of the three internal patterns examined, the honeycomb pattern had the greatest impact resistance, as shown in Figure 4.7. When sampled using the honeycomb structure, the impact strength increased steadily between 1.9 kJ/m^2 at 75% infill and a 0.1 mm thickness, and 2.6 kJ/m^2 at 95% infill and a 0.3 mm thickness. The honeycomb structure was cellular, which was presumably better at shock absorption due to a more even distribution of the impact load throughout the interior geometry. The triangular design gave medium outcomes with a range between 1.7 and 2.2 kJ/m^2, which indicated good energy dissipation, but it was not very efficient compared to the honeycomb design. Conversely, the brick pattern had the lowest impact strengths of 1.4 to 1.8 kJ/m^2. This weakness may be due to the fact that it has weaker load transfer pathways and a more fragmented structure that might have some local-stress concentration during impact.

Regarding the infill density, there was a noticeable trend of rising impact strength and energy intake with the increase in the infill density between 75 and 95%. In the honeycomb series, the impact strength rose to 2.6 kJ/m^2 at 95% infill compared to 1.9 kJ/m^2 at 75% infill. The triangular and brick patterns also displayed this trend, but the improvement was minimal. These density increments probably improved the internal bonding and reduced the void content, which led to improved energy transfer. Like other mechanical properties, the increase in impact strength also started to level off at the higher density range, indicating a diminishing marginal benefit in constructing the structure stiffer and less able to deform to absorb shock energy.

The influence of layer thickness had a generally positive trend too, as in the honeycomb and brick designs. Greater impact strength (S3 at 2.6 kJ/m^2) was achieved with thicker layers (0.3 mm), presumably because interlayer fusion was better and because few interfacial defects existed. These parameters are critical in impact situations where

TABLE 4.6
Impact Test Results for All Experimental Samples

Sample ID	Pattern	Infill density	Layer thickness	Impact strength (kJ/m²)	Impact energy (J)
S1	Honeycomb	75%	0.1 mm	1.9	2.6
S2	Honeycomb	85%	0.2 mm	2.3	3.1
S3	Honeycomb	95%	0.3 mm	2.6	3.6
S4	Triangular	75%	0.2 mm	1.7	2.3
S5	Triangular	85%	0.3 mm	2.0	2.8
S6	Triangular	95%	0.1 mm	2.2	2.9
S7	Brick	75%	0.3 mm	1.4	2.0
S8	Brick	85%	0.1 mm	1.6	2.2
S9	Brick	95%	0.2 mm	1.8	2.4

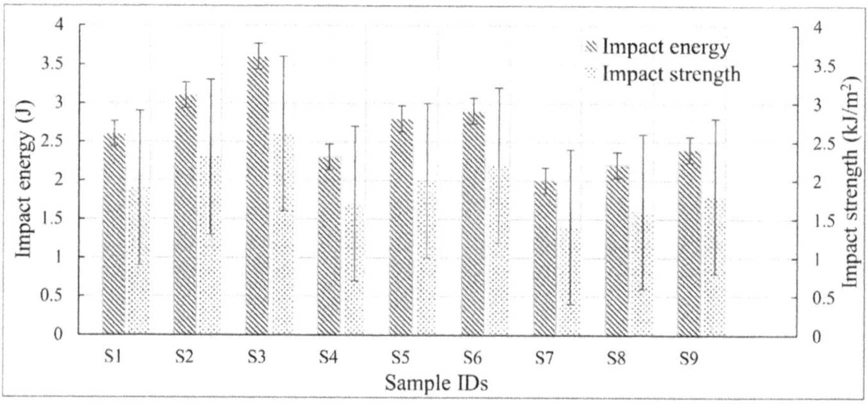

FIGURE 4.7 Effect of printing process parameter on impact strength.

the delamination or weak layer bonding may easily cause crack propagation and fracture. Nevertheless, the trend was less obvious in the triangular pattern: although the sample of a 0.3 mm layer (S5) was reasonable and equal or even better in certain cases (2.0 kJ/m^2), the samples of 0.10 mm (S6) and 0.20 mm (S4) were no worse in this case. This indicates that pattern geometry/layer height interaction could have a more unusual effect on impact resistance than in tensile or flexural applications.

4.6 CONCLUSIONS

The ABS+CF specimens were developed using the FFF technique, and a combined effect of pattern, density, and layer thickness was observed on the structural properties. The results give the following conclusions:

- The infill pattern was significant in the mechanical behavior of all the tests. The honeycomb pattern had the maximum tensile strength and impact resistance, with a maximum of 42.1 MPa UTS and 2.6 kJ/m^2 impact strength, and the triangular pattern had the highest flexural strength of 72.0 MPa. In comparison, the maximum value of the brick pattern was always lower than the maximum value of the other tests: 35.9 MPa (tensile), 37.0 MPa (flexural), and 1.8 kJ/m^2 (impact).
- The infill density was generally positively correlated with all the mechanical properties. The tensile strength increased by 16.3% (36.2 to 42.1 MPa), flexural strength followed by 22% (59.0 to 72.0 MPa), and impact strength by 37% (1.9 to 2.6 kJ/m^2) with an increase in infill density (75 to 95%) with the optimum settings. Nevertheless, the enhancements stopped at a point lower than 85; this means that 85% infill can lead to the most favorable trade-off between strength and material consumption.
- The thickness of the patterns' layers had a positive impact on tensile strength to 0.3 mm, flexural strength to 11.0 MPa and impact strength to 0.7 kJ/m^2. Increased interlayer adhesion and structural integrity was achieved with increased layer thickness, but eventually, dimensional accuracy might be lost.
- Not only was the optimum arrangement of each property a little different, but a general tendency was observed in that high infill density (85–95%), moderate-to-thick layers (0.2–0.3 mm), and judicious choice of pattern depended on the type of load. The honeycomb design, with 95% infill, 0.3 mm layers, is most suitable in tensile and impact applications, while the triangular design, with 85% infill, 0.3 mm layers in flexural applications, is optimal.
- The findings are useful in offering guidance on optimizing the parameters of the FFF process on ABS+CF composites. The most effective design is a combination of a honeycomb or triangular infill, 85 to 95% density and 0.3 mm layer thickness, as an engineer would want to balance mechanical performance, depending upon the tensile, flexural, or impact resistance requirements.

These results are helpful in the attempts of the professionals to enhance the mechanical performance of FDM-produced ABS+CF parts. The determined optimal parameter set provides a useful guideline in applications that need balanced mechanical traits. A comprehensive knowledge of the effects of the parameters can help in adapting the print environment to the needs of a product's particular performance.

REFERENCES

Agrawal, A. P., Kumar, V., Kumar, J., Paramasivam, P., Dhanasekaran, S., & Prasad, L. (2023). An investigation of combined effect of infill pattern, density, and layer thickness on mechanical properties of 3D printed ABS by fused filament fabrication. *Heliyon*, *9*(6), e16531. https://doi.org/10.1016/j.heliyon.2023.e16531

Akhoundi, B., & Behravesh, A. H. (2019). Effect of filling pattern on the tensile and flexural mechanical properties of FDM 3D printed products. *Experimental Mechanics*, *59*(6), 883–897. https://doi.org/10.1007/s11340-018-00467-y

Alafaghani, A., Qattawi, A., Alrawi, B., & Guzman, A. (2017). Experimental optimization of fused deposition modelling processing parameters: A design-for-manufacturing approach. *Procedia Manufacturing*, *10*, 791–803. https://doi.org/10.1016/J.PROMFG.2017.07.079

Algarni, M. (2021). The influence of raster angle and moisture content on the mechanical properties of PLA parts produced by fused deposition modeling. *Polymers*, *13*(2), 237. https://doi.org/10.3390/POLYM13020237

Attaran, M. (2017). The rise of 3-D printing: The advantages of additive manufacturing over traditional manufacturing. *Business Horizons*, *60*(5), 677–688. https://doi.org/10.1016/J.BUSHOR.2017.05.011

Caminero, M. Á., Chacón, J. M., García-Plaza, E., Núñez, P. J., Reverte, J. M., & Becar, J. P. (2019). Additive manufacturing of PLA-based composites using fused filament fabrication: Effect of graphene nanoplatelet reinforcement on mechanical properties, dimensional accuracy and texture. *Polymers*, *11*(5), 799. https://doi.org/10.3390/POLYM11050799

Cantrell, J. T., Rohde, S., Damiani, D., Gurnani, R., DiSandro, L., Anton, J., Young, A., Jerez, A., Steinbach, D., Kroese, C., & Ifju, P. G. (2017). Experimental characterization of the mechanical properties of 3D-printed ABS and polycarbonate parts. *Rapid Prototyping Journal*, *23*(4), 811–824. https://doi.org/10.1108/RPJ-03-2016-0042

Carneiro, O. S., Silva, A. F., & Gomes, R. (2015). Fused deposition modeling with polypropylene. *Materials & Design*, *83*, 768–776. https://doi.org/10.1016/J.MATDES.2015.06.053

Chacón, J. M., Caminero, M. A., García-Plaza, E., & Núñez, P. J. (2017). Additive manufacturing of PLA structures using fused deposition modelling: Effect of process parameters on mechanical properties and their optimal selection. *Materials & Design*, *124*, 143–157. https://doi.org/10.1016/J.MATDES.2017.03.065

Dizon, J. R. C., Espera, A. H., Chen, Q., & Advincula, R. C. (2018). Mechanical characterization of 3D-printed polymers. *Additive Manufacturing*, *20*, 44–67. https://doi.org/10.1016/J.ADDMA.2017.12.002

Fernandes, J., Deus, A. M., Reis, L., Vaz, M. F., & Leite, M. (2018). Study of the influence of 3D printing parameters on the mechanical properties of PLA. *Proceedings of the International Conference on Progress in Additive Manufacturing*, *2018-May*, 547–552. https://doi.org/10.25341/D4988C

Gibson, I., Rosen, D., Stucker, B., & Khorasani, M. (2020). *Additive Manufacturing Technologies*, 3rd edition, 1–675. https://doi.org/10.1007/978-3-030-56127-7

Goh, G. D., Yap, Y. L., Tan, H. K. J., Sing, S. L., Goh, G. L., & Yeong, W. Y. (2020). Process–structure–properties in polymer additive manufacturing via material extrusion: A review. *Critical Reviews in Solid State and Materials Sciences*, *45*(2), 113–133. https://doi.org/10.1080/10408436.2018.1549977

Kumar, N., Jain, P. K., Tandon, P., & Pandey, P. M. (2019). Investigations on the melt flow behaviour of aluminium filled ABS polymer composite for the extrusion-based additive manufacturing process. *International Journal of Materials and Product Technology*, *59*(3), 194–211. https://doi.org/10.1504/IJMPT.2019.102931

Kuznetsov, V. E., Solonin, A. N., Urzhumtsev, O. D., Schilling, R., & Tavitov, A. G. (2018). Strength of PLA components fabricated with fused deposition technology using a desktop 3D printer as a function of geometrical parameters of the process. *Polymers*, *10*(3), 313. https://doi.org/10.3390/POLYM10030313

Liu, X., Zhang, M., Li, S., Si, L., Peng, J., & Hu, Y. (2017). Mechanical property parametric appraisal of fused deposition modeling parts based on the gray Taguchi method.

International Journal of Advanced Manufacturing Technology, *89*(5–8), 2387–2397. https://doi.org/10.1007/S00170-016-9263-3
Lubombo, C., & Huneault, M. A. (2018). Effect of infill patterns on the mechanical performance of lightweight 3D-printed cellular PLA parts. *Materials Today Communications*, *17*, 214–228. https://doi.org/10.1016/J.MTCOMM.2018.09.017
Mishra, P. K., Senthil, P., Adarsh, S., & Anoop, M. S. (2021). An investigation to study the combined effect of different infill pattern and infill density on the impact strength of 3D printed polylactic acid parts. *Composites Communications*, *24*, 100605. https://doi.org/10.1016/J.COCO.2020.100605
Ngo, T. D., Kashani, A., Imbalzano, G., Nguyen, K. T. Q., & Hui, D. (2018). Additive manufacturing (3D printing): A review of materials, methods, applications and challenges. *Composites Part B: Engineering*, *143*, 172–196. https://doi.org/10.1016/J.COMPOSITESB.2018.02.012
Popescu, D., Zapciu, A., Amza, C., Baciu, F., & Marinescu, R. (2018). FDM process parameters influence over the mechanical properties of polymer specimens: A review. *Polymer Testing*, *69*, 157–166. https://doi.org/10.1016/J.POLYMERTESTING.2018.05.020
Raja, S., Agrawal, A. P., P Patil, P., Thimothy, P., Capangpangan, R. Y., Singhal, P., & Wotango, M. T. (2022). Optimization of 3D printing process parameters of polylactic acid filament based on the mechanical test. *International Journal of Chemical Engineering*, *2022*(1), 5830869. https://doi.org/10.1155/2022/5830869
Rajpurohit, S. R., & Dave, H. K. (2018). Effect of process parameters on tensile strength of FDM printed PLA part. *Rapid Prototyping Journal*, *24*(8), 1317–1324. https://doi.org/10.1108/RPJ-06-2017-0134
Rankouhi, B., Javadpour, S., Delfanian, F., & Letcher, T. (2016). Failure analysis and mechanical characterization of 3D printed ABS with respect to layer thickness and orientation. *Journal of Failure Analysis and Prevention*, *16*(3), 467–481. https://doi.org/10.1007/S11668-016-0113-2
Rodríguez, J. F., Thomas, J. P., & Renaud, J. E. (2001). Mechanical behavior of acrylonitrile butadiene styrene (ABS) fused deposition materials. Experimental investigation. *Rapid Prototyping Journal*, *7*(3), 148–158. https://doi.org/10.1108/13552540110395547
Sood, A. K., Ohdar, R. K., & Mahapatra, S. S. (2010). Parametric appraisal of mechanical property of fused deposition modelling processed parts. *Materials & Design*, *31*(1), 287–295. https://doi.org/10.1016/J.MATDES.2009.06.016
Srivastava, P., Arumugam, A. B., Agrawal, A. P., & Ntumba, Z. N. (2025). Development of hemp fiber reinforced PLA composites for sustainable 3D printing: Mechanical and microstructural properties. *Journal of Natural Fibers*, *22*(1), 2558214. https://doi.org/10.1080/15440478.2025.2558214
Tanveer, M. Q., Suhaib, M., & Haleem, A. (2021). A retrospective investigation of mechanical and physical properties of ABS specimen developed by manual injection moulding and fused deposition modelling. *Lecture Notes in Mechanical Engineering*, 965–978. https://doi.org/10.1007/978-981-15-5463-6_85
Tofail, S. A. M., Koumoulos, E. P., Bandyopadhyay, A., Bose, S., O'Donoghue, L., & Charitidis, C. (2018). Additive manufacturing: Scientific and technological challenges, market uptake and opportunities. *Materials Today*, *21*(1), 22–37. https://doi.org/10.1016/J.MATTOD.2017.07.001
Torrado, A. R., & Roberson, D. A. (2016). Failure analysis and anisotropy evaluation of 3D-printed tensile test specimens of different geometries and print raster patterns. *Journal*

of Failure Analysis and Prevention, *16*(1), 154–164. https://doi.org/10.1007/S11668-016-0067-4

Wang, X., Jiang, M., Zhou, Z., Gou, J., & Hui, D. (2017). 3D printing of polymer matrix composites: A review and prospective. *Composites Part B: Engineering*, *110*, 442–458. https://doi.org/10.1016/J.COMPOSITESB.2016.11.034

Wickramasinghe, S., Do, T., & Tran, P. (2020). FDM-based 3D printing of polymer and associated composite: A review on mechanical properties, defects and treatments. *Polymers*, *12*(7), 1529. https://doi.org/10.3390/POLYM12071529

5 Modeling and Simulation of Mechanical Processes

Utpal Sen, Mandeep Pathak, Borad M Barkachary, and Anil Kumar Deepati

5.1 INTRODUCTION

Additive manufacturing (AM), also known as rapid prototyping, is tremendously popular due to its short production time, freedom of design, and improved control over process variables, such as layer thickness and printing speed, and flow rate of molten filament. In all rapid prototyping processes or methods, 3D printing has emerged not only as a method for generating prototypes, but also as a means of producing end products within a production process. This method not only reduces material wastage, but also allows makers or manufacturers to introduce modifications easily. There are various methods of AM, also known as 3D printing, including laminated object manufacturing (LOM), selective laser sintering (SLS), stereolithography (SLA), and direct energy deposition (DED). Fused deposition modeling (FDM) is particularly popular and widely used due to its use of polylactic acid (PLA), a thermoplastic polymer, which is easy to use and has a lower cost for daily life, including day-to-day tasks and operations. In addition to PLA, examples of material used for FDM include polyphenylsulfone (PPSU), acrylonitrile butadiene styrene (ABS), polycarbonate (PC), polyetherimide (PEI), polyvinyl acetate (PVA), and poly-caprolactone (PCL). Among these materials, PLA has an edge over the others due to its affordability (low cost) and biodegradability (Byrley et al., 2019; Chua et al., 2017; Miron et al., 2017). The selection of the appropriate material for 3D printing is important to produce the required component or part. Additionally, two factors—infill pattern or structures and density—significantly influence the amount of material used, the strength and integrity of the final printed part, as well as the production time (Harpool, 2016). The amount or quantity of material that fills the internal part's volume is called the infill density, whereas infill pattern denotes the geometric pattern the infilled volume material follows (Baich, 2016). Sucuoglu et al. (2018) conducted research on the effect of infill patterns produced by 3D printing on the structural strength of 3D printed components. Research focused on developing a hybrid infill design to improve 3D printing efficiency (Sucuoglu et al., 2018), and another study by the authors (Bogrecki et al., 2019) showed that the hexagonal infill pattern provides superior structural strength. A study was also conducted on how different filament materials (PLA, ABS, nylon) and slicing software (Ultimaker Cura, PrusaSlicer, Repetier-Host) affect the performance of fused filament fabrication (FFF) 3D-printed parts. Using ASTM D638

DOI: 10.1201/9781003643531-5

Type 1 dimensions, parts were assessed for print time accuracy, dimensional accuracy, and surface quality. Using ABS with Ultimaker Cura produced the smoothest surface (2.792 μm), while using PLA with Repetier-Host achieved the best dimensional accuracy (0.03%) (Maideen et al., 2025). A study was conducted to evaluate the tensile strength of 3D-printed PLA parts using ASTM D638 (Type IV specimens). As it is a biodegradable material prevalently used in FDM 3D printing, PLA was selected for the investigation. Specimens were modelled in SolidWorks and printed on a Makerbot 3D printer. After proper conditioning, the samples underwent tensile testing using an Autograph AG-15 Universal Testing Machine (UTM). ANSYS software was employed to determine the maximum deformation and tensile strength. By using CAD models of the species (Anand Kumar & Shivraj Narayan, 2019), research was conducted to analyze the impact of strain rate on the tensile characteristics of PLA, which is fabricated by the atomic diffusion manufacturing (ADM) method. In this study, ASTM D638 Type V specimens were used, printed on a 3D printer. These specimens underwent three distinct strain rates: 2 mm per minute, 0.8 mm per minute and 20 mm per minute, respectively. To evaluate the ultimate (maximum) tensile strain, digital image correlation (DIC) was employed (Eltahir et al., 2024). Bhuiyan (2025) investigated the mechanical properties of AM polymers in his thesis and described the anisotropy and interlayer bonding of the components. Garone et al. (2017) explored the use of a desktop 3D printer to print a microfluidic prototype, demonstrating the applicability of desktop printers. Ngo et al. (2018) provided a brief introduction to the methods, materials, applicability and challenges of 3D printing in their review article. Ray et al. (2025) explored various methods to enhance the surface properties of Kevlar-reinforced composites produced by additive manufacturing. Rezaei et al. (2024) described how nanoscale reinforcements influence the mechanical properties of 3D-printed parts, and also performed simulations on PLA and reinforced PLA. Roberts (2020) focused on the development of a rotational printhead system for fiber control during extrusion, demonstrating the flexibility of 3D printing systems.

Rogelj et al. (2025) investigated the tensile properties of carbon fiber composites with PLA as the matrix material, accounting for thermal management during printing. Stojković et al. (2023) conducted an experimental study on the impact of layer height, and after printing, annealing was performed as post-processing, which improved mechanical properties. Yavas et al. (2021) studied the fracture behavior of 3D-printed carbon fiber-reinforced polymer composites. The failure mechanisms were analyzed under loading conditions to determine the reliability of parts manufactured through additive manufacturing.

5.2 METHODOLOGY

5.2.1 Preparation of the CAD Model

To begin, research was conducted by reading current research literature on 3D printing to build a solid foundation for the research investigation. This first step was very important to chart the course of the project. From here, a systematic research methodology was used to start the project, starting by first selecting a standard specimen for testing. An ASTM D638 Type I specimen was selected as shown in Figure 5.1, which

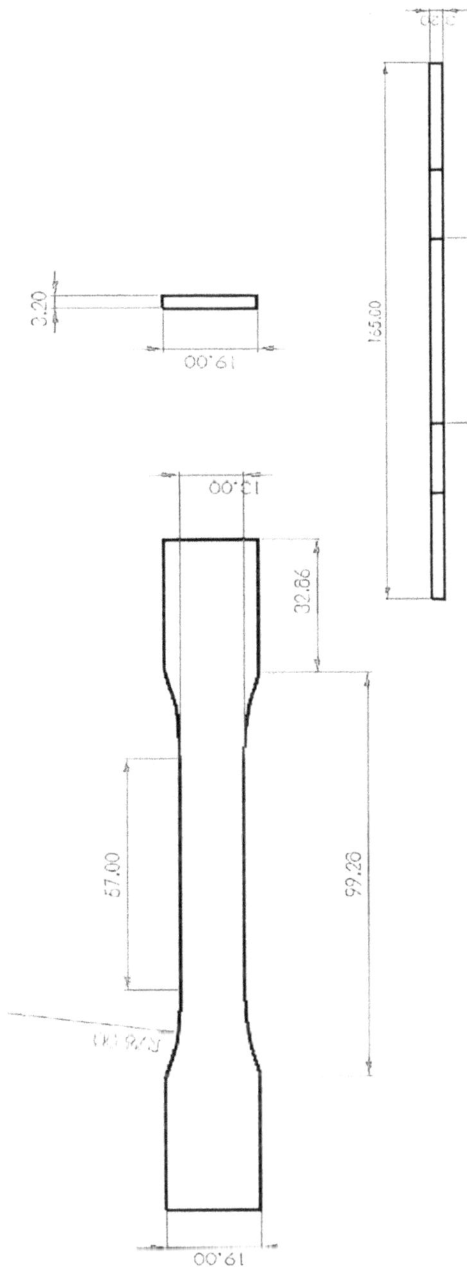

FIGURE 5.1 Tensile specimen ASTM D638 (Type I).

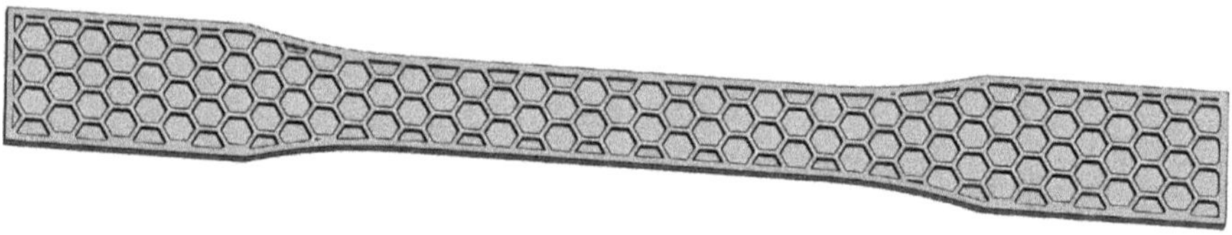

FIGURE 5.2 Infill pattern made in SolidWorks.

is frequently used in tensile testing process for material properties. This is because very specific dimensional parameters are specified for an ASTM D638 Type I test specimen, which is generally used for mechanical testing, such as tensile testing.

The major measurements are the span of the narrow portion of the specimen, which is 13 mm, and the length of the narrow portion of the specimen, which has a value of 57 mm. The minimum overall width of the standard specimen is 19 mm, and the minimum overall length of the specimen is 165 mm. The gauge length of the standard specimen is given as 50 mm. The thickness is 3.20 mm. The radius of the fillet of the standard sample is 76 mm. The distance between the handles of the specimen is 115 mm.

SolidWorks is very popular computer-aided design (CAD) software, used to design specimens to make a digital model for 3D printing. After the design phase, the CAD models are transferred to slicing software. The slicing software that was used in this study was PrusaSlicer, which converts or transforms a CAD file in STL format to a code file format. This G code instructs the 3D printer to move in three dimensions by giving the coordinates of the positions it should use, and it then extrudes the fused polymer material layer by layer. ANSYS is an excellent tool for looking at how materials behave mechanically, but it cannot read the G-code file, thus the technique needs to be changed. To overcome this issue, the method is modified. Instead of relying on the slicer-generated G-code file, the specimen is recreated in SolidWorks, as shown in Figure 5.2, this time incorporating key features such as grid and honeycomb infill patterns with infill densities of 20%, 40%, 60%, and 80% directly into the model.

5.2.2 The Slicing Process

After generating the model in SolidWorks, the resulting CAD file, designed as a fully solid structure with a 100% infill and no internal hollowness, is converted to STL format, and then imported into PrusaSlicer for the slicing process. Within PrusaSlicer, the infill density can be customized independently of the original CAD model. Although the model is initially created as a solid, the slicer interprets it as a volume to be filled according to the user-defined infill percentage. For example, when selecting infill densities of 20%, 40%, 60%, or 80% in the slicer, varying percentages or amounts of internal hollowness are generated within the structure. Grid and honeycomb patterns were generated for this particular study, as shown in Figure 5.3 and Figure 5.4. The printed part remains fully solid for a specimen with 100% infill.

FIGURE 5.3 Slicing of honeycomb pattern in PrusaSlicer.

FIGURE 5.4 Slicing of grid pattern in PrusaSlicer.

5.2.3 Preparation for 3D Printing

5.2.3.1 3D Printer Used

For this study, an Olivetti 3D desktop printer was used, shown in Figure 5.5. It was developed by an Italian tech company, Olivetti. As this is a desk-sized printer, it is designed for small labs and personal use for the rapid prototyping of models and small components, offering good precision and accuracy. Its compactness makes it suitable for use in limited spaces. The materials that can be used in this printer are PLA, ABS, and nylon. The features of this 3D printer are also mentioned in Table 5.1.

5.2.3.2 Material Used

PLA is a widely used material in 3D printing, valued for its affordability, ease of use, and eco-friendliness. PLA is a biodegradable thermoplastic that is suitable for both professionals and novices, since it is derived or obtained from renewable resources

FIGURE 5.5 Olivetti 3D desktop printer.

TABLE 5.1
Features of the 3D Printer

General features	
Technology	FDM
Printing area	200 × 200 × 200 [mm]
Printing volume	About 8 [liters]
Body	The machine body is anodised aluminum 30 × 30 radius. The case is in polycarbonate
Extruder	Single
Maximum speed	Up to 200 [mm/s]
Nozzles	Availability of various diameter nozzles, from 0.30–0.80 mm (default nozzle: 0.40 mm)
Printing management	Direct printing from an SD card is possible without a PC
Control panel	Integrated LCD control panel

TABLE 5.2
Properties of PLA Material

Property	Value	Unit
Density	1250	[kg/m^3]
Shear modulus	1241	[MPa]
Bulk modulus	5227.3	[MPa]
Poission's ratio	0.39	-
Young's modulus	3450	[MPa]

such as sugarcane or cornstarch. It is compatible with indoor printing due to its minimal odor, lack of a heated platform, and low melting point (180–220°C). PLA comes in a range of colors and unique blends, such as flexible and composite alternatives that include wood or metal mixed in. It has good strength, rigidity, and surface polish, so it is good for models that do not need to be very flexible. It does not hold up to heat as well as ABS and other materials, and it may change shape when it becomes hot. The material properties are shown in Table 5.2.

5.2.3.3 3D Printing of Specimens

In this investigation, test samples were produced through 3D printing as shown in Figure 5.6, with two distinct infill patterns: honeycomb and grid, shown in Figure 5.8 and Figure 5.7. For each pattern, specimens were printed with different infill densities as follows: 20%, 40%, 60%, and 80%. A fully solid specimen was also produced with a 100% infill. Using honeycomb and grid patterns ensures that it is easy to compare different infill geometries. This introduces variations so that a relative study can be done on both infill patterns.

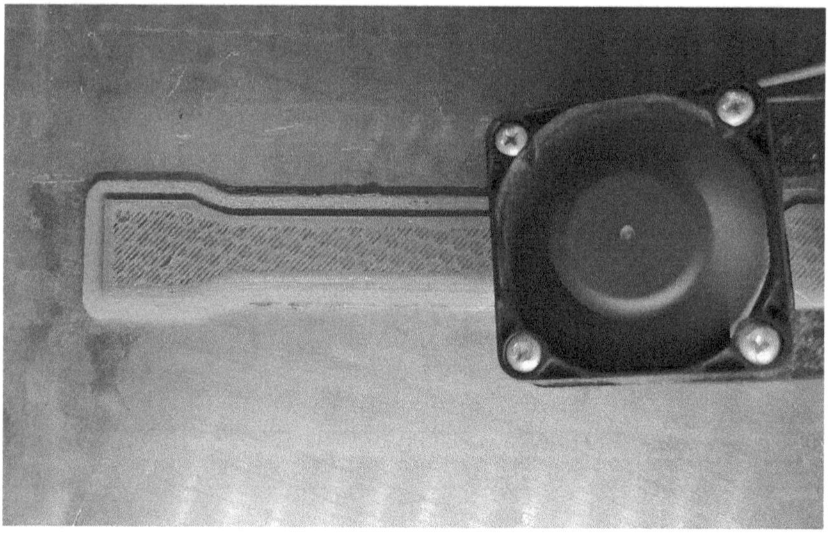

FIGURE 5.6 3D printing.

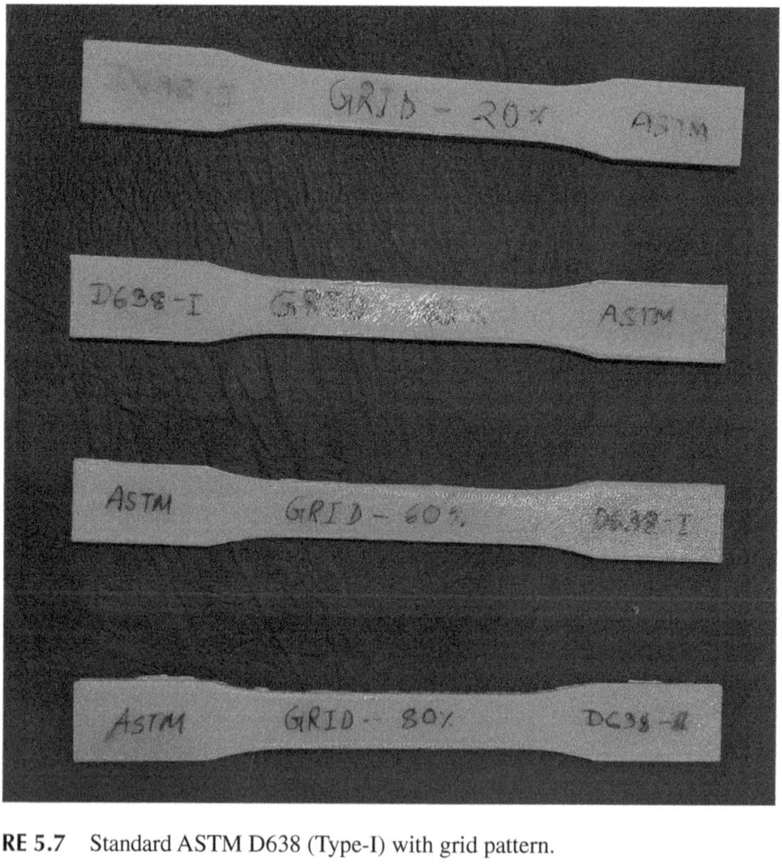

FIGURE 5.7 Standard ASTM D638 (Type-I) with grid pattern.

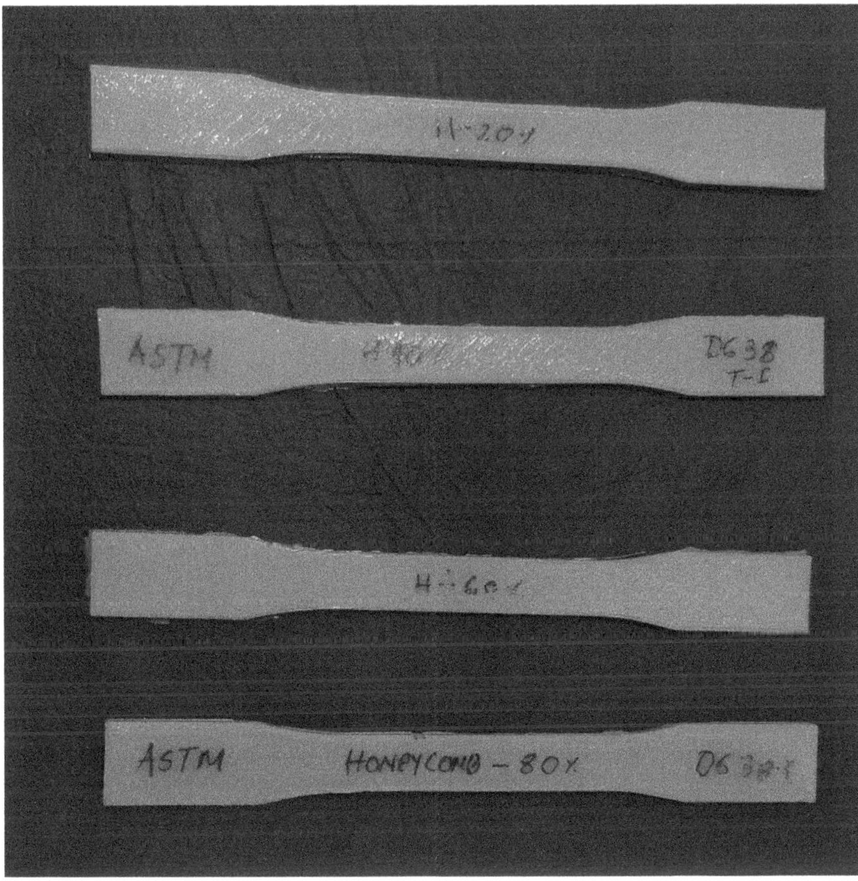

FIGURE 5.8 Standard ASTM D638 (Type-I) with honeycomb infill.

5.2.4 Tensile Testing

Mechanical tests on all specimens were conducted using a Model MTS Acumen UTM, shown in Figure 5.9. Specifications of the UTM are shown in Table 5.3. The specimens were fabricated with grid and honeycomb infills at 20%, 40%, 60%, 80%, and 100% densities after the completion of the 3D printing process. Tensile testing was performed to determine and examine the mechanical performance of the specimens shown in Figure 5.10, Figure 5.11 and Figure 5.12, specifically by examining loading bifurcations, failure mode, elongation, and tensile strength under uniaxial loading. The experiment was standardized to allow the loading of the specimens to be kept consistent so that the performance properties of specimens with different infills could be appropriately compared and assessed. The work in tensile testing for this study provided evidence on the mechanical performance of 3D-printed specimens based on the type of infill and density. Observations showed that specimens with higher infill densities exhibited better tensile strength and stiffness.

FIGURE 5.9 MTS Acumen UTM.

Source: **North East Centre for Biological Sciences and Healthcare, IIT Guwahati.**

TABLE 5.3
Specifications of UTM

Model no.	MTS Acumen 12
Static load	8.5 [kN]
Dynamic load	12 [kN]
Range	0–12 [kN]
Applicable test	Static, fatigue, three-point bending, and torsion tests are performed.

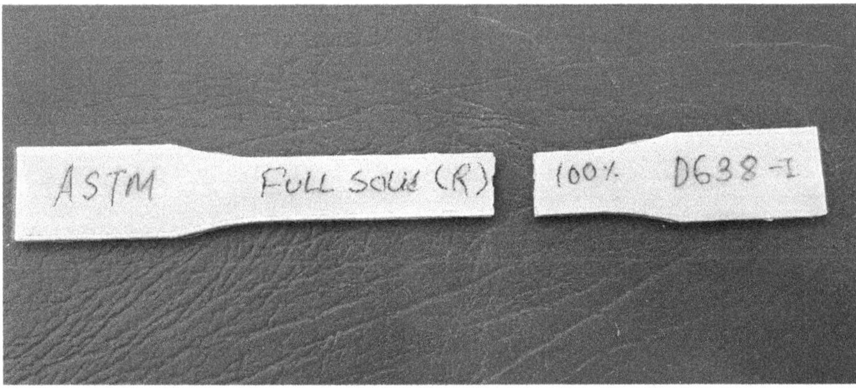

FIGURE 5.10 Full solid specimen after testing.

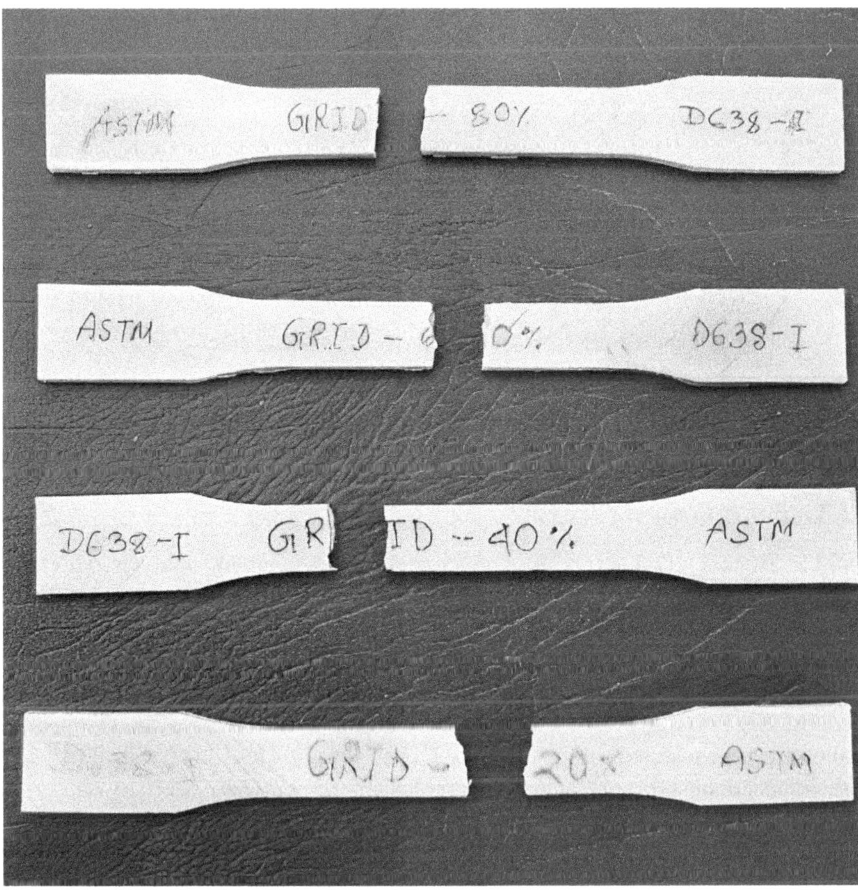

FIGURE 5.11 Grid infill specimens after testing (80%, 60%, 40%, 20%).

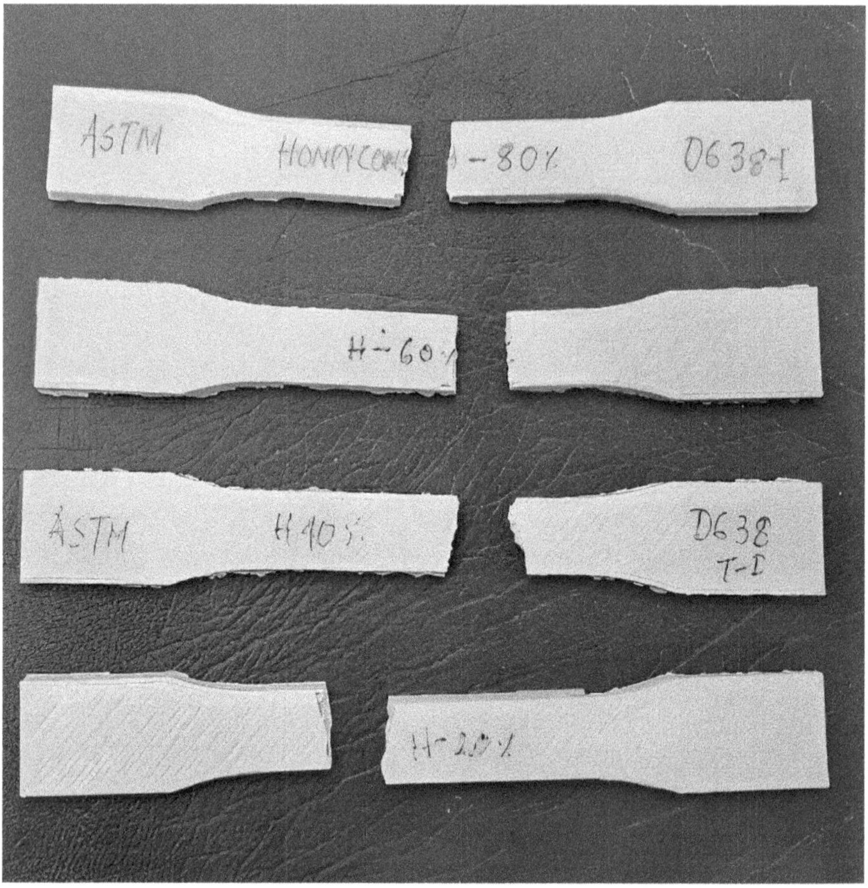

FIGURE 5.12 Honeycomb infill specimens after testing (80%, 60%, 40%, 20%).

5.2.5 Finite Element Analysis

ANSYS Student 2025 RI is used to conduct finite element analysis, shown in Figure 5.13. All the CAD models of the samples with grid and honeycomb infill structures with densities of 20%, 40%, 60% and 80% were processed in the software. As there was no pre-existing information on PLA in the material library, the values of density, Young's modulus (also known as the modulus of elasticity), yield strength, and Poisson's ratio were manually added to the library. The static structural option was selected for the analysis. Constraints and tensile loading conditions were imposed on the specimens or samples in the software itself to obtain the desired results, such as strain, stress, and deformation, as well as failure points.

The computational method provided results that were understandable in view, while still being detailed, including the overall deformation, Von Mises stress, and strain concentration of the specimens. There were interesting results from the finite element analysis related to the experimental trends seen from the tensile test, since

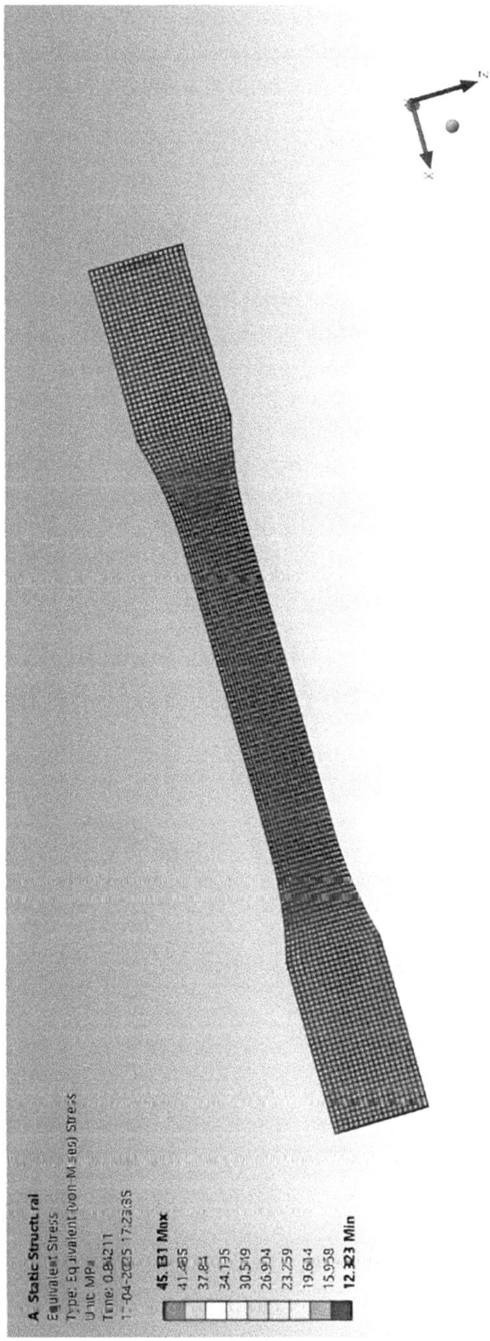

FIGURE 5.13 Analysis in ANSYS software

the specimens that had a higher infill percentage showed less deformation and a more uniform stress distribution, whereas specimens with a lower infill showed a more flexible specimen with larger areas of stress concentration. As a function of the internal shape difference between a honeycomb and grid pattern, the finite element analysis illustrated how the infill pattern changes the load route, as well as the manner of the location where failures started.

5.3 RESULTS AND DISCUSSION

5.3.1 Results Obtained from the Experiment

5.3.1.1 Experimental Stress-strain Curve Analysis of 3D-printed Specimens

The solid specimen with 100% infill density provided the highest stress before breaking (32.8316 MPa). The second highest was the honeycomb 80% (H80) (32.83002 MPa), followed by grid 80% (G80) (26.90822 MPa). Honeycomb and grid 20% (H20 and G20) had a very low value of 19–24 MPa.

The solid infill 100% specimen showed the maximum strain percentage, which indicates the highest ductility compared to the grid pattern. The honeycomb pattern showed relatively high ductility. The solid 100% infill showed the highest toughness of all. The toughness then decreased by grid 80%, honeycomb 80% and grid 60% (G80, H60, and G60), respectively. The honeycomb 20% (H20) and grid 20% (G20) had the lowest toughness of them all.

From the graph shown in Figure 5.14, it is evident that the solid specimen exhibited the highest stress and strain before breaking. The infill pattern, with increased

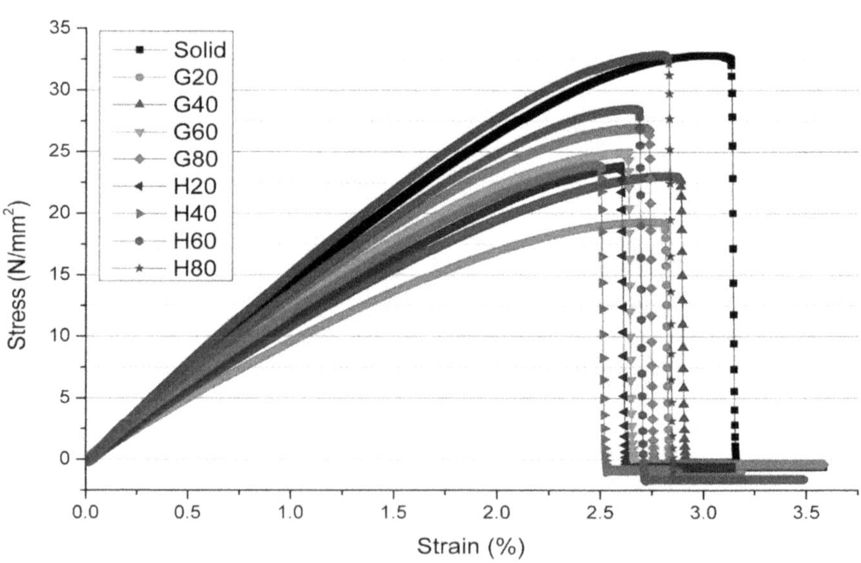

FIGURE 5.14 Stress-strain curves obtained from physical testing.

TABLE 5.4
Data Obtained from Physical Testing

Infill type	Density	UTS (MPa)	Strain at break (%)	Toughness
Solid	100%	30.74062	2.973684	VH
Honeycomb (H80)	80%	30.66876	2.761199	VH
Grid (G80)	80%	25.04938	2.659152	H
Honeycomb (H60)	60%	26.59448	2.631652	H
Grid (G60)	60%	22.82287	2.61272	M
Honeycomb (H40)	40%	21,77677	2.492442	M
Honeycomb (H20)	20%	21.7332	2.59443	L–M
Grid (G40)	40%	18.51518	2.719254	L–M
Grid (G20)	20%	17.16178	2.637734	L

• VH = Very high; M = Moderate; VL = Very low; H = High; L = Low

intensity, reached a greater peak value of both strain and stress. Additionally, it is observed that the honeycomb pattern exhibited relatively good performance compared to the grid pattern. This was due to its uniform stress distribution and load-handling capacity. The data obtained from the physical testing of all the specimens is shown in Table 5.4.

5.3.2 Results Obtained from Simulation

5.3.2.1 Stress-strain Curve Analysis of Simulated CAD Specimens

The simulation was conducted in ANSYS Student R1, a popular simulation software that is well known for its simulation capabilities, and static structural analysis. It was employed with appropriate load conditions to mimic an experimental environment. The data obtained from the simulation is given in Table 5.5. In the case of the simulation, the solid specimen with 100% infill density also provided the highest stress before breaking (30.74062 MPa) and recorded the highest strain at 2.973684% at the point of breaking, shown in Figure 5.15. The second highest was honeycomb 80% (H80), which had the highest or peak stress of 30.66876 MPa and strain—2.761199% at the time of breaking, while grid 80% (G80) provided a peak stress of 25.04938 MPa and a strain of 2.659152% at the time of breaking. Here, it was also observed that, with increasing infill density, the strain and stress values at breakage increased. The honeycomb and grid 20% (H20 and G20) provided the lowest value on breaking. This trend in mechanical performance with respect to infill density and pattern revealed a discernible relationship in terms of ductility, overall structural robustness, and stiffness.

In observations, it is found that the honeycomb specimens with different infill densities exhibited greater mechanical performance than the grid specimens at the same densities. Due to uniform material distribution, the solid 100% sample remained the strongest of all.

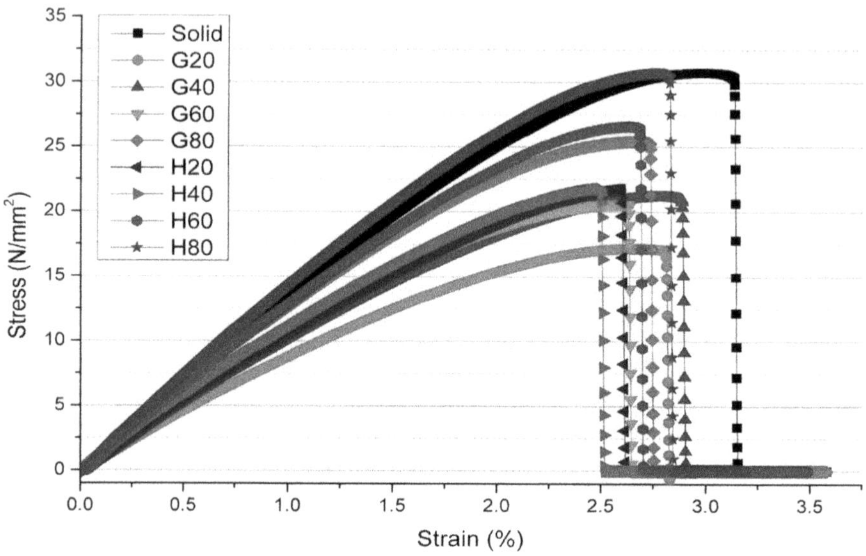

FIGURE 5.15 Stress-strain curves obtained from simulation.

TABLE 5.5
Data Obtained from the Simulation

Infill type	Density (%)	UTS (MPa)	Strain at break (%)	Toughness
Solid 100	100%	32.8316	3.038	VH
Honeycomb (H80)	80%	32.83002	2.787456	H
Grid (G80)	80%	26.90822	2.688406	H
Honeycomb (H60)	60%	28.48169	2.646213	M
Grid (G60)	60%	25.03534	2.624372	M
Honeycomb (H40)	40%	23.94383	2.49538	L
Grid (G40)	40%	22.91998	2.814313	L
Honeycomb (H20)	20%	23.79424	2.59443	VL
Grid (G20)	20%	19.28114	2.73858	VL

• VH = Very high; M = Moderate; VL = Very low; H = High; L = Low

5.3.3 Comparison Between Experimental and Simulated Results

Although experimental and simulated results show some relations and similarity, there are some differences because the simulated specimens do not take account of the impact of the environmental and physical world, such as room temperature. The moisture content of the environment in which the experiment was conducted will always be variable because simulated and actual conditions, or real-world conditions, are not entirely perfectly similar. As discussed earlier, a solid 100% sample exhibits a stress (peak) of 32.8316 MPa, while simulations show a stress of 30.74062 MPa, which is slightly less than the experimental result. This trend is also observed for other specimens. Subsequently, this trend is also observed in the strain at the point of the break. The stress-strain curves show some differences in shape. The experiments show the nuanced and complex beginning of deformation that occurs under load. The curves turned out smoothly from the elastic regime to visible strain hardening and then dropped off at peak stress. For the simulation, the shape of the curves looked closer to ideal with constant slopes, with no complexity or variability in the non-linear zone. This meant that there was no sign of localized necking or crack propagation in the actual world.

The important difference in the results was the spread or clumping of the curves. Experimentally, the curves were shown to be spread out or varied because every print was slightly different due to the variability in the 3D printing process, for example, adhesion between the layers, and different nozzle temperatures, whereas the simulation results produced very grouped results, and showed an inability of the simulation to capture the natural variability and uncertainty.

When comparing some individual samples, solid 100% infill emerges as the strongest in both differences of each set of data. However, the experimental data showed higher stress and lower ductility than the simulated data, but the simulation produced slightly lower strength and better ductility. In low-density samples, such as grid and honeycomb 20% (G20 and H20), both results confirmed the expected drop in mechanical strength due to high porosity and lack of internal reinforcement, with simulations slightly underestimating the stress, but successfully mirroring the overall trend. For grid and honeycomb 40% (G40 and H40), the experimental stress declined more noticeably, whereas the simulation maintained a smoother profile with less abrupt changes for Indus grid and honeycomb 60% (G60 and H60). There was a significant difference between the experiment and a similar experiment that obtained curved shot non-linearity, This indicates improper fluctuations of stress. This occurred due to imperfections within the printed part. The grid and honeycomb 80% (G80 and H80) samples showed the highest performance in both experimental and simulation aspects. This referred to the fact that the higher the density, the higher the load-bearing capacity and ductility.

Again, there was a major difference in the modulus of elasticity. Experiments showed a sudden breaking due to the imperfections of the 3D printed parts, such as poor interlayer adhesion. However, in the simulation, the stress values were low because of the absence of imperfections.

5.3.4 Similarity Between Experimental and Simulated Results

There are clear similarities in the behavioral patterns of both the grid and honeycomb series, with different infrared densities. Firstly, both showed the same trend, namely that when the infill density increases, the stress and strain at breakage increase accordingly. The lowest infill density showed the lowest value, while the highest density showed the highest value of both stress and strain at breaking. This trend is evident from the order: G20 < G40 < G60 < G80 and H20 < H40 < H60 < H80.

TABLE 5.6
Observation Table

Aspect	Observation (common for grid and honeycomb)
Stress-density relationship	Stress increases with increasing density
Stress-strain curve shape	Elastic → plateau → densification (similar pattern)
Deformation behavior	Lower density → easy deformation; higher density → stiffer
Solid sample behavior	Highest stress and stiffness for both series

Secondly, in both cases, the graph looks similar. Both the graphs showed a linear elastic region, a yielding or plateau region, and a densification region. This pattern is characteristic of cellular or foam-like structures, regardless of whether the internal geometry is grid-based or honeycomb-based.

Furthermore, the effect of density on deformation behavior is consistent in both infill patterns. Lower density specimens tend to deform more easily, resulting in lower stress values, but higher strain capacities. In contrast, higher density specimens offer greater resistance to deformation, thereby achieving higher stress values, but lower strains at break. Finally, the solid (100%) sample, being common to both comparative frameworks, stands out with the highest stiffness, the maximum stress-bearing capability, and the least deformation. This behavior suggests a more brittle or limited ductility nature compared to the more compliant and energy-absorbing behavior of infilled samples. Observations are also tabulated in Table 5.6.

5.4 CONCLUSION

This research provides an idea of how infill patterns, infill structures and infill densities can affect the mechanical performance and efficiency of the structure of the printed component using FDM by simulation and experimental tensile testing. It is seen that infill configuration plays a significant role in the mechanical properties of the printed parts. Among the tested patterns, the honeycomb pattern with 80% infill density (H80) demonstrated the best balance between mechanical strength and material efficiency. Moreover, this configuration achieved the highest stress of 30.67 and 32.83 MPa in the simulation and experimental setups, respectively, with maximum strain at breaking, 2.76 and 2.8% for the honeycomb 80% infilled specimen (H80), which makes it structurally strong, as well as lightweight.

The solid 100% specimen exhibited the highest tensile properties, specifically 32.84 MPa in the experiment, and 30.75 MPa in the simulation, which is suitable for heavy loading applications. However, as the material usage is high, it is less flexible and weighs more; again, grid and honeycomb 20% (G20 and H20) low-density infills showed very unsuitable performance for their design. A comparison between the experimental and simulated results is necessary to perform the comparison, which revealed imperfections like microcracks, edge issues, crack propagation, and stress concentration that affect the structure under load in real-world conditions.

This demonstrates the necessity of checking to obtain accurate and reliable mechanical properties. In conclusion, the study shows the significance of design choices in FDM 3D printing. Carefully selecting the required infill pattern and density can highly affect the structural strength and material efficiency of a specific design. These are highly useful in manufacturing, especially in aerospace, automotive, consumer goods, and biomedical engineering industries and applications.

REFERENCES

Anand Kumar, S., Shivraj Narayan, Y. (2019). Tensile Testing and Evaluation of 3D-Printed PLA Specimens as per ASTM D638 Type IV Standard. In: Chandrasekhar, U., Yang, LJ., Gowthaman, S. (eds) *Innovative Design, Analysis and Development Practices in Aerospace and Automotive Engineering (I-DAD 2018). Lecture Notes in Mechanical Engineering*. Springer, Singapore. https://doi.org/10.1007/978-981-13-2718-6_9

Baich, L. (2016). Impact of Infill Design on Mechanical Strength and Production Cost in Material Extrusion Based Additive Manufacturing‖, Master of Science in Engineering in the Industrial and Systems Engineering Program [Master's Thesis], Youngstown State University, Youngstown.

Bhuiyan, M.Z.H. (2025). Influence of Process Parameters on the Mechanical Performance of 3D Printed ABS: A Comparative Study of Experimental, Numerical, and Statistical Analysis (Order No. 32173806). (3239181777) [Master's Thesis]. www.proquest.com/dissertations-theses/influence-process-parameters-on-mechanical/docview/3239181777/se-2

Bogrekcı, İ., Demırcıoglu, P., Sucuoglu, H.S., Turhanlar, O. (2019). The Effect of the Infill Type and Density on Hardness of 3d Printed Parts. *International Journal of 3D Printing Technologies and Digital Industry*, 3, 3, 212–219. https://dergipark.org.tr/en/download/article-file/910652

Byrley, P., George, B.J., Boyes, W.K., Rogers, K., (2019). Particle Emissions from Fused Deposition Modelling 3D Printers: Evaluation and Meta-Analysis. *Science of Total Environment*, 655, 395–407. https://doi.org/10.1016/j.scitotenv.2018.11.070

Chua, C.K., Wong, C.H., Yeong, W.Y. (2017). *Standards, Quality Control, and Measurement Sciences in 3D Printing and Additive Manufacturing*. Academic Press.

Eltahir, S.A., Gomaa, R., Yilmaz, C. (2024). Effect of Strain Rate on the Tensile Properties of 3D–Printed PLA Specimens with Fused Deposition Modelling. *Journal of Advanced Manufacturing*, 5, 2, 37–46. https://doi.org/10.14744/ytu.jame.2024.00006

Eltahir, S., Gomaa, R., Yilmaz, C. (2024). Effect of Strain Rate on the Tensile Properties of 3D–Printed PLA Specimens with Fused Deposition Modelling. *Journal of Advances in Manufacturing Engineering*, 5, 2, 37–46. https://doi.org/10.14744/ytu.jame.2024.00006

Garone, P. (2017). Exploring the Use of Desktop 3D Printing for Microfluidics Prototyping, Thesis in the Field of Biotechnology for the Degree of Master of Liberal Arts in Extension Studies [Master's Thesis], Harvard University, USA.

Harpool, T.D. (2016). Observing the Effects of Infill Shapes on the Tensile Characteristics of 3D Printed Plastic Parts [Bachelor's Thesis] Master of Science on the Department of Mechanical Engineering and the Faculty of the Graduate School of Wichita State, Wichita.

Maideen, N.C., Hyie, K.M., Abdullah, N.S., Mohd Fohimi, N.A., Yusoff, H., Nahar, A.H.M., Sahudin, S. (2025). An Optimum Combination of Filament Material and Slicing Software in Improving the Manufacturing Performance of 3D-printed Parts. *Scientific Research Journal*, 22, 135–149. https://doi.org/10.24191/srj.v22is.13289

Miron, V., Ferrandiz, S., Juarez, D., Mengual, A., (2017). Manufacturing and Characterisation of 3D Printer Filament using Tailoring Materials‖. *Manufacturing Engineering Society International Conference (MESIC), Procedia Manufacturing*, 13, 888–894, Vigo, Spain. https://doi.org/10.1016/j.promfg.2017.09.151

Ngo, T.D., Kashani, A., Imbalzano, G., Nguyen, K.T.Q., Hui, D. (2018). Additive Manufacturing (3D Printing): A Review of Materials, Methods, Applications and Challenges‖. *Composites Part B*, 143, 172–196. https://doi.org/10.1016/j.compositesb.2018.02.012

Ray, N.C., Saha, R.K., Mollah, Md E., Rakib, S., Ali, Y. (2025). Enhancing Mechanical and Surface Properties of 3D-Printed Kevlar-Reinforced ABS/PLA Composites through FDM Process. *Hybrid Advances*, 11, 100510, ISSN 2773-207X. https://doi.org/10.1016/j.hybadv.2025.100510.

Rezaei, A., Izadi, R., Fantuzzi, N. (2024). A Hierarchical Nano to Micro Scale Modelling of 3D Printed Nano-Reinforced Polylactic Acid: Micropolar Modelling and Molecular Dynamics Simulation. *Nanomaterials*, 14, 1113. https://doi.org/10.3390/nano14131113

Roberts, M.D. (2020). Design and Implementation of Rotational Print Heads to Control Fiber Orientation in Material Extrusion Additive Manufacturing. [Master's Thesis], University of Tennessee. https://trace.tennessee.edu/utk_gradthes/6097

Rogelj, A., Liović, D., Hozdić, E., Franulović, M., Mijović, B. (2025). Influence of Cooling Lubricants and Structural Parameters on the Tensile Properties of FFF 3D-Printed PLA and PLA/Carbon Fiber Composites. *Polymers*, 17, 1797. https://doi.org/10.3390/polym17131797

Stojković, J.R., Turudija, R., Vitković, N., Górski, F., Păcurar, A., Pleşa, A., Ianoşi-Andreeva-Dimitrova, A., Păcurar, R. (2023). An Experimental Study on the Impact of Layer Height and Annealing Parameters on the Tensile Strength and Dimensional Accuracy of FDM 3D Printed Parts. *Materials*, 16, 4574. https://doi.org/10.3390/ma16134574

Sucuoglu, H.S., Bogrekci, I., Demircioglu, P., Gultekin, A. (2018). The Effect of Three-Dimensional Printed Infill Pattern on Structural Strength‖. *El-Cezeri Journal of Science and Engineering*, 5, 3, 785–796. https://doi.org/10.31202/ecjse.423915

Sucuoglu, H.S., Bogrekci, I., Demircioglu, P., Turhanlar, O. (2018). Development of Hybrid Pattern for Three-Dimensional Printing Optimization. *Sigma Journal of Engineering and Natural Sciences*, 36, 3, 667–675. https://dergipark.org.tr/en/download/article-file/2036259

Yavas, D., Zhang, Z., Liu, Q., Wu, D. (2021). Fracture Behavior of 3D Printed Carbon Fiber-Reinforced Polymer Composites. *Composites Science and Technology*, 208, 108741, ISSN 0266-3538, https://doi.org/10.1016/j.compscitech.2021.108741.

6 Tribological Performance of Additive Manufactured Materials Under Sliding Applications

Ribhu Kundu, Akant Kumar Singh, Anant Krishan Pun, Deepak Kapila, Manvendra Yadav, and Mohamed Sukri Mat Ali

6.1 INTRODUCTION

Generally, the conversion of a starting material into a product via the application of energy is termed the manufacturing process (Batista et al., 2021). In recent studies, the significant influence of wear and friction on global energy consumption has been emphasized. Nearly 23% of the world's total energy use is associated with tribological contacts, with 20% consumed in overcoming friction, and an additional 3% utilized for remanufacturing worn components and replacing equipment owing to wear-related failures (González-Hernández et al., 2022). A critical issue in the industry is controlling wear and friction in moving machine parts. Due to the sliding, rolling, or impacting of two solid surfaces against one another, material loss or damage takes place (Njim et al., 2024). The wastage of material happens owing to the birth of additive manufacturing (AM), in which the final object is prepared by the addition of material (Singh & Bharti, 2021). The lightweight AM parts may be rendered by the topology optimization of AM. They have the same structural performance as traditional components when featuring a diminished weight simultaneously (Sola et al., 2024). Modern AM technology initiates with stereolithography (SLA) technology, which starts with the processing of a polymer (hardening) selectively in a layer-by-layer fashion (Sapkota et al., 2024). Layer-by-layer deposition takes place in AM Unlike subtractive manufacturing methodologies, the amount of waste material can be limited, thus drastically mitigating overall costs and conserving resources (Shah et al., 2022). In 3D printing, solid ground curing, direct digital manufacturing, and layer manufacturing, AM technology is very popular among its users (Mahmood et al., 2022). 3D printing is one of the major types of AM. 3D printing is utilized to analyze the mechanical properties of parts fabricated with different advanced technologies (Portoacă et al., 2023). Polymers act as a fundamental material in 3D

DOI: 10.1201/9781003643531-6

printing, thereby offering versatility to accommodate a broad range of final product properties, as well as permitting customization to meet the creator's specific needs (Iftekar et al., 2023). A general thermoplastic polymer that is made by the polymerization of acrylonitrile and styrene in the existence of polybutadiene is known as acrylonitrile butadiene styrene (ABS) (Srinivasan et al., 2020).

ABS, manufactured by AM technology, renders parts with a certain geometric accuracy and also provides mechanical or surface properties (Alami et al., 2023). Particularly, the fused filament fabrication (FFF) process, also named fused deposition modeling (FDM), has become essential with a high economical and low environmental impact (Daminabo et al., 2020). Furthermore, when analogized with other AM technologies, FDM is more popular owing to its ease of utilization and low cost. In this advanced AM technology, biodegradable composites are produced by the combination of polylactic acid (PLA) with natural fibers (Aida et al., 2021). But, the FDM approach has significant drawbacks, like inadequate strength for load-bearing applications, fragile performance, and limited suitability for fully functional components (Keshavamurthy et al., 2021). Hence, in the prevailing research studies, different solutions were developed for improvement.

Kumar et al. (2021) examined the covering wear characteristics of AM printed specimens with materials such as aluminum, nickel, steel, ABS, and PLA. In the research, the three stages of the wear test, namely the laboratory test, component simulation test, and in-service test, were declared. For PLA, the maximum wear rate was obtained as 1300 μm at 90 orientations. Also, the wear test, utilizing a ball on the disc, depicted that, for all loads, wear volume loss was less in horizontally built specimens when analogized with vertically built specimens. Nevertheless, in the research study, specific tests were only focused on the wear behavior analysis, thus resulting in a poor generalizability problem. Conversely, Pant et al. (2020) analyzed the variation of process parameters, namely layer thickness, orientation, and extruder temperature. The FDM approach was utilized in the research framework. By utilizing the Taguchi L9 orthogonal array design of experiments, the process parameters were optimized. Furthermore, for rigorous validation and optimization processes, the ANalysis Of VAriance (ANOVA) and artificial neural network (ANN) techniques were utilized. As per the outcomes, the most dominating factor for the design experiments was the orientation of the printing. However, the ANN model did not appropriately predict the performance curve values due to the lack of data. Likewise, Chhabra et al. (2023) propounded the analysis of optimal parameters of FFF-3D printing utilizing carbon-reinforced nylon composite material to improve sliding wear performance. For training, predicting, and optimizing the sliding wear performance of the fabricated parts, the Genetic Algorithm-Adaptive Neuro-Fuzzy Inference System (GA-ANFIS) was utilized. The experimental analysis depicted that the minimum sliding wear rate (i.e. 0.008636 mm^3/m) was obtained at a layer height of 0.28 mm. But, in the genetic algorithm, the imbalance issue happened in the initialization of the population position, thereby affecting the composite material's performance.

Keshavamurthy et al. (2023) recommended the mechanical, as well as wear studies of boron nitride (BN)-reinforced polymer composites that were established through 3D printing technology. The research incorporated various percentages of

BN into the PLA matrix, especially at 5% and 10% by weight, for the evaluation. The compounding process involved the blending of BN and PLA, followed by the creation of composite filaments measuring 1.75 mm in diameter. The specific amount of BN was only considered in the research model, thus degrading the reliability of the research study. Contrarily, Albahkali et al. (2023) concentrated on the development of an ANN model utilizing ANFIS to predict the wear volume of PLA composites under different conditions. Also, to understand the tribological outcomes, and assess the load-bearing ability of the PLA composites, the research framework evaluated the mechanical properties and finite element model. When contrasted with experimental outcomes, the ANFIS model had an average percentage error of less than 0.01%. Nevertheless, the ANFIS approach had a high computational cost problem, thus degrading the research framework's performance.

Likewise, Guenther et al. (2021) researched the tribological performance of American Iron and Steel Institute H-13 (AISI H13) steel samples. The samples were tested under four diverse conditions, namely polished, ground and polished, as-printed, and laser-surface-textured (LST) with round dimples. The experiments of the samples were tested against bearing steel disks in a unidirectional pin-on-disk tribometer. The study result depicted that, among the four different conditions, grinding and polishing had a low friction coefficient. But, the tribological performance investigation by four different conditions was not enough. Slapnik et al. (2020) aimed to evaluate the tribological performance of 3D printed parts produced by utilizing vat photopolymerization (VPP). By using Charpy impact tests, tensile tests, dynamic mechanical analysis (DMA), and Shore D, the thermal and mechanical properties of the material were tested in the research model. The experimental analysis displayed that the modified resins diminished the coefficient of friction (52%) and wear rate (92%) when analogized with unmodified photocurable resins. Conversely, Zakaulla et al. (2021) propounded the ANN model-centric mechanical properties' prediction for a polyether ether ketone (PEEK) composite reinforced with titanium and graphene powder. By utilizing input vectors as the weight of PEEK and reinforcement material, the multi-layer perception feed-forward neural network was constructed in the ANN model. When compared to the experimental measurements, the presented ANN model obtained satisfactory outcomes in the experimental evaluation. However, the traditional multi-layer perception feed-forward had an over-fitting problem, thereby affecting the model's performance.

To test the material, different techniques have been utilized in some research. For instance, Nataraja et al. (2024) introduced a unique approach to improve the tribological properties of the Al-12% weight of Si alloy matrix material. The material was manufactured with altering percentages of micro-sized ZrO_2 reinforcements, especially 0.5%, 1%, and 3% weight. To declare the behavior of the material under different operational conditions, the result of the research study was optimized and validated by utilizing response surface methodology (RSM) techniques and an ANOVA table, respectively. According to the experimental study, the wear rate was decreased by up to 3% in the weight of reinforcements, which was increased owing to reinforcement agglomeration. The study did not evaluate the hardness values centered on material property changes, thereby degrading the framework's performance. Contrarily, Ergene et al. (2025) analyzed the influences of seawater on PLA and carbon fiber-reinforced PLA manufactured at diverse infill rates (i.e. 20%, 60%, and 100%). In the research study, pin-on-disc wear with a height of 20 mm and a diameter of 10 mm was tested in the SolidWorks 2023 design program as per ASTM G99. The study findings exposed that, for the coefficient of friction values of the PLA and carbon fiber-reinforced PLA parts, wear volume loss result was attained, which altered between 0.37 and 0.75. In the study, the layer thickness of the material was not estimated, thus affecting the reliability of the research. Prevailing research did not evaluate the material wear behavior under different orientations of the material parts. Thus, horizontally and vertically printed ABS specimens using pin-on-disk testing and ANN modeling are presented in this study.

6.2 OBJECTIVE OF THE STUDY

The objectives of the study presented in this chapter are as follows:

- Consider the ABS specimens for fabrication testing using FDM with two diverse orientations of ABS specimen designs.
- Generate wear test values by utilizing the pin-on-disk tribometer.
- Predict the wear coefficient rate and wear loss utilizing ANN modeling and the wear rate estimation equation, correspondingly.
- Evaluate the performance of ANN modeling and the wear rate estimation equation centered on the wear coefficient rate and wear loss.

6.3 MATERIALS AND METHODS

In the research study, an analysis of the wear characteristics of 3D-printed ABS specimens fabricated in two different build orientations (horizontal and vertical) is presented (see Figure 6.1). Owing to variations in interlayer adhesion and anisotropy introduced during the FDM process, the orientation of the specimens influences the tribological and mechanical properties of the specimens.

Figure 6.2 presents a structural diagram of the proposed research method.

6.3.1 Materials

In the research study, the ABS filament specimens are utilized for the fabrication test. By using FDM, fabrication testing is performed. FDM is a popular 3D printing technology that works by melting filament and extruding it into a printing bed via a nozzle to form layers. Furthermore, FDM has good wear resistance and mechanical strength. The specimens are designed in the form of cylindrical pins with a length of 40 mm and a diameter of 8 mm, appropriate for pin-on-disk wear testing. An infill density of 15% is selected for all printed samples to ensure uniformity throughout all tests. The specimens are printed in two different orientations—horizontal and vertical—to investigate the impact of the build direction on wear performance.

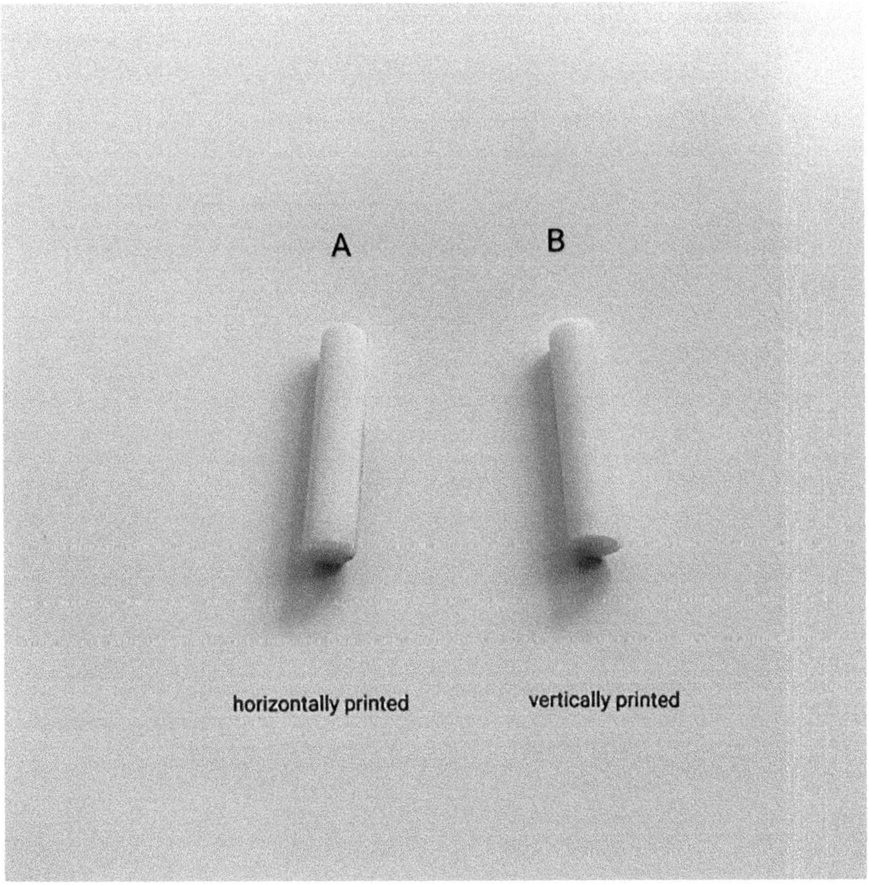

FIGURE 6.1 ABS specimens fabricated in horizontal and vertical orientations.

The FDM process parameters are displayed in Table 6.1. The print quality, surface finish, and mechanical properties of the printed parts are significantly influenced by the FDM process parameters, thus affecting their wear performance in tribological tests.

6.3.2 Methods

The presented research study uses a standard pin-on-disk tribometer for the wear test. The disk material utilized during testing is standardized as hardened steel. All experiments are carried out in ambient environmental conditions. As per the L27 orthogonal array design, the rotational speeds, applied loads, and sliding distances alter. However, other conditions, like disk material and the environment, remain fixed. The wear volume and corresponding wear coefficient are recorded for every single test. Further, they are utilized for ANN model training and validation.

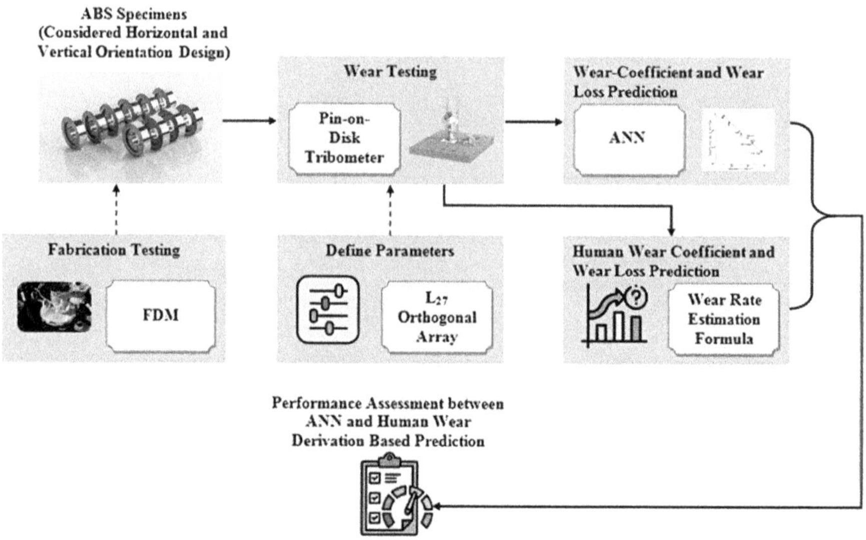

FIGURE 6.2 Structural diagram for the proposed research methodology.

TABLE 6.1
FDM Parameters

Parameters	Values
Nozzle temperature	230–240°C
Bed temperature	100–110°C
Printing speed	40–100 mm/s
Filament diameter	1.75 mm

Initially, by utilizing the ANN model, the data obtained from the wear test is trained. The normal load, sliding distance, sliding speed, and printing orientation are given as input to the ANN model. The wear coefficient and wear volume are taken as the output from the ANN model. Here, ($\aleph_{SS}$) represents the combination of input parameters. The ANN model has numerous advantages, like the potential to learn from input data, generalize well to unseen data, and easily adapt to new information. Figure 6.3 illustrates the structural diagram of the ANN model.

Three layers—the hidden layer, input layer, and output layer—are included in the ANN model. Every single network is a three-layer feed-forward ANN with a single hidden layer of 10 neurons. For training, the research model uses the Levenberg-Marquardt learning algorithm, owing to its fast convergence and high accuracy in wear modeling. In the hidden layer, a log-sigmoid (logsig) activation function is utilized. Moreover, for the output neuron, linear (purelin) activation is used. The input layer passes the scaled input information to the next layer. Here, (λ) signifies the

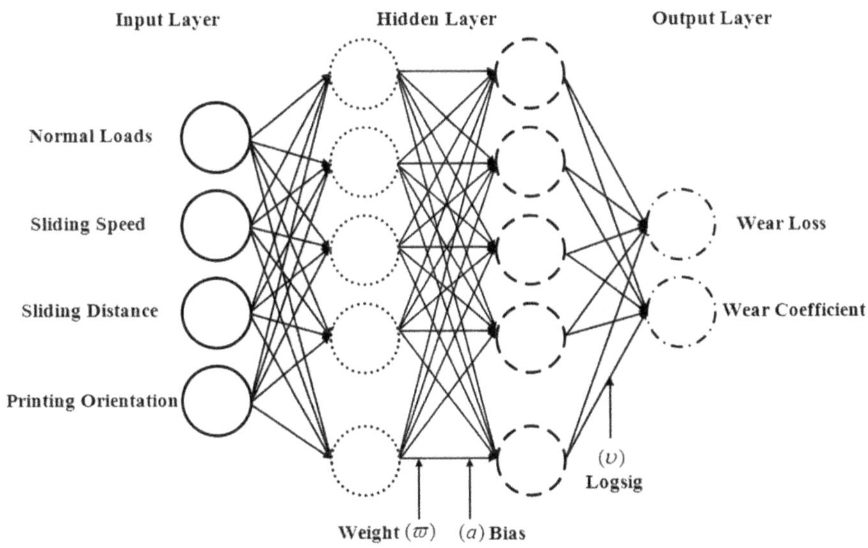

FIGURE 6.3 Architecture for ANN approach.

outcome of the input layer. Afterwards, every single neuron performs a weighted sum of inputs with bias in the hidden layer. To obtain the result, an activation function is applied. The mathematical expression for the hidden layer operation is depicted in Equation (1) as:

$$\beta_j = \sum \varpi . \aleph_{ss} + a \tag{1}$$

$$\upsilon = \alpha\left(\beta_j\right) \tag{2}$$

$$\alpha\left(\beta_j\right) = \frac{1}{1 + e^{-\beta_j}} \tag{3}$$

Where, the weight value is notated as ϖ, the bias value of the neuron is implied as a, the jth neuron in the first hidden layer is indicated as β_j, the exponential function is exemplified as e, and the output of the neuron in the hidden layer is denoted as υ. The outcome of the wear test is decided in the output layer. Generally, the output layer has a linear activation function. The mathematical derivation of the output layer is provided in Equation (4) as:

$$\xi = \sum \varpi \cdot \upsilon + a \tag{4}$$

Here, the outcome of the output layer that provides the final wear coefficient outcome is represented as ξ.

6.3.3 Orthogonal Array

The Taughi methodology is utilized in the research study to design the experimental matrix for wear testing. The L27 orthogonal array design is used, in particular, in this research study. The L27 orthogonal array design is highly effective in diminishing the number of experiments needed to study multiple factors when ensuring sufficient data for analysis. To systematically study the effects of normal load (2 N, 11 N, and 20 N), sliding speed (150 RPM, 425 RPM, and 700 RPM), and sliding distance (800 m, 2150 m, and 3500 m) on wear, the Taguchi L27 orthogonal design (3 factors × 3 levels) is utilized. The use of an L27 array mitigates the number of experiments needed, while maintaining orthogonality. All tests are carried out under dry sliding conditions.

The detailed design values of the Taughi L27 orthogonal array design for wear testing are depicted in Table 6.2, ensuring that each level of each factor is tested equally and in combination with all levels of the other factors.

6.3.4 Experimental Setup

Primarily, the wear testing is done by handling the pin-on-disk test. The wear coefficient values are initially computed from the obtained wear test values by utilizing the wear equation. For every single run, the wear volume loss is evaluated by considering the specific wear rate (also known as the wear coefficient) of ABS. The volume of material removed per unit load per unit sliding distance (with unit mm^3/Nm) is referred to as the specific wear rate ($\omega_{r'}$). Hence, the wear volume ($\wp$) is calculated for a given test with sliding distance ($\Im$) and normal load (η), and articulated as:

$$\wp = \omega_{r'} \times \eta \times \wp \tag{5}$$

For wear evaluation, it is assumed that lower wear resistance is slightly better displayed by horizontally printed ABS specimens than vertically printed ones. This is attributed to enhanced interlayer adhesion and a smoother contact surface in the horizontal orientation, resulting in greater material removal during sliding contact. In other words, the horizontal specimens are expected to exhibit a higher specific wear rate (i.e. larger ($\omega_{r'}$) and ($\wp$)) under identical conditions when compared to the vertical specimens. This assumption will be reflected in the estimated wear values for every single orientation.

Then, by utilizing the trained ANN model, the wear coefficient of the ABS specimens in the horizontal and vertical directions is automatically predicted centered on the pin-on-disk test values. After that, to validate the model's accuracy,

TABLE 6.2
Detail Matrix Values Taguchi L27 Orthogonal Array Design for Wear Testing

Run	Normal load (N)	Sliding speed (RPM)	Sliding distance (m)
1	2	150	800
2	2	150	250
3	2	150	3500
4	2	425	800
5	2	425	250
6	2	425	3500
7	2	700	800
8	2	700	250
9	2	700	3500
10	11	150	800
11	11	150	250
12	11	150	3500
13	11	425	800
14	11	425	250
15	11	425	3500
16	11	700	800
17	11	700	250
18	11	700	3500
19	20	150	800
20	20	150	250
21	20	150	3500
22	20	425	800
23	20	425	250
24	20	425	3500
25	20	700	800
26	20	700	250
27	20	700	3500

the derived values in Equation (5) are analogized with the ANN-predicted values. By utilizing a regression plot (predicted vs. actual values), a mean squared error (MSE) performance plot, and an error histogram, the accuracy and reliability of the model are assessed to visualize the distribution of prediction errors. The comparison of the research establishes the ANN's reliability in approximating the wear behavior of ABS under diverse tribological conditions.

6.4 RESULTS

Here, the performance of the ANN model-centric wear coefficient rate and wear loss estimation is investigated. The presented research is implemented on the working platform of MATLAB.

6.4.1 Dataset Description

The ANN model's performance is analyzed with the help of data generated from the wear test utilizing a pin-on-disk tribometer. The dataset, which includes 27 cases for each orientation, is partitioned into training, validation, and testing subsets in the ratio of 70:15:15. Specifically, to train the network, 19 randomly selected runs are utilized, and the remaining eight runs are split evenly for validation (i.e. to monitor overfitting issues) and final testing.

6.4.2 Performance Analysis

In this section, the experimental validation of the human wear test and the ANN wear test is investigated.

The performance evaluation of the horizontal direction-based ABS specimen design is depicted in Table 6.3 and Table 6.4. A total of 27 test cases are included in this table. Every single test case contained different combinations of load and

TABLE 6.3
Performance Analysis of Human Weight Loss and ANN Weight Loss Analysis in the Horizontal Direction

Run	Load (N)	Speed (RPM)	Distance (m)	Wear volume (mm³)	Wear coefficient (mm³/N·m)
1	2	150	800	6235	3.8978
2	2	150	2150	16772	3.9034
3	2	150	3500	27970	3.9963
4	2	425	800	6911	4.3194
5	2	425	2150	18246	4.2446
6	2	425	3500	30234	4.3208
7	2	700	800	7548	4.7180
8	2	700	2150	20842	4.8369
9	2	700	3500	34181	4.8548
10	11	150	800	30681	4.8001
11	11	150	2150	84116	4.8836
12	11	150	3500	137654	5.2398
13	11	425	800	34388	6.2659
14	11	425	2150	93977	6.1186
15	11	425	3500	154488	6.0611
16	11	700	800	39678	7.2214
17	11	700	2150	107159	7.1773
18	11	700	3500	174270	7.1147
19	20	150	800	47176	2.9485
20	20	150	2150	127412	2.9596
21	20	150	3500	207835	2.9692
22	20	425	800	53493	3.3410

TABLE 6.3 (Continued)
Performance Analysis of Human Weight Loss and ANN Weight Loss Analysis in the Horizontal Direction

Run	Load (N)	Speed (RPM)	Distance (m)	Wear volume (mm^3)	Wear coefficient ($mm^3/N \cdot m$)
23	20	425	2150	144796	3.3577
24	20	425	3500	236502	3.3769
25	20	700	800	56864	3.5537
26	20	700	2150	155390	3.6088
27	20	700	3500	248379	3.6517

TABLE 6.4
Performance Analysis of Human Wear Coefficient and ANN Wear Coefficient Analysis in the Horizontal Direction

Run	Estimated K ($mm^3/N \cdot m$)	ANN predicted K ($mm^3/N \cdot m$)	Error (%)
1	3.8978	3.8463	1.32%
2	3.9034	3.9823	2.02%
3	3.9963	3.9124	2.10%
4	4.3194	4.2846	0.80%
5	4.2446	4.3155	1.67%
6	4.3208	4.3823	1.42%
7	4.7180	4.6581	1.27%
8	4.8369	4.7572	1.65%
9	4.8548	4.7820	1.50%
10	4.8001	4.8659	1.37%
11	4.8836	4.8050	1.61%
12	5.2398	5.1730	1.27%
13	6.2659	6.3491	1.33%
14	6.1186	6.0477	1.16%
15	6.0611	6.1447	1.38%
16	7.2214	7.1520	0.96%
17	7.1773	7.0815	1.33%
18	7.1147	7.1964	1.15%
19	2.9485	2.9830	1.17%
20	2.9596	2.9124	1.59%
21	2.9692	2.9327	1.23%
22	3.3410	3.3138	0.81%
23	3.3577	3.4041	1.38%
24	3.3769	3.4205	1.29%
25	3.5537	3.5002	1.51%
26	3.6088	3.5579	1.41%
27	3.6517	3.7210	1.90%

sliding distance, with corresponding weight loss by human measurement and ANN modeling provided in milligrams (mg). According to the analysis, the lower load scenarios (i.e. 2N and 11N) had the strongest agreement with the ANN-based prediction compared to experimental values. Numerous test cases obtained an accuracy of above 85%. Table 6.3 and Table 6.4 depicted that, when contrasted with human experimentation, the ANN model efficiently predicted weight loss owing to sliding wear in the horizontal direction of the ABS specimens.

In Figure 6.4, the experimental wear coefficient values derived by humans are analogized with those predicted by the ANN modeling across 27 different test cases under the horizontal orientation of ABS specimens. For the ANN prediction model, the wear coefficient was higher for low-load scenarios (i.e. test cases 1 to 9). When the load was increased, the wear coefficient decreased. The largest difference was presented between human prediction and ANN model prediction for test cases 3, 5, and 8. Overall, the investigation depicted that, in most cases, the ANN-predicted wear coefficients were closely matched to the experimental values.

The comparative analysis between the human-measured weight loss and the ANN-predicted weight loss value of ABS specimens under different test conditions in the vertical sliding direction is illustrated in Table 6.5 and Table 6.6. The prediction model's accuracy remained between 80 and 90% for the loads in test cases 1 to 9. The ANN prediction model continuously rendered a reliable outcome with minor deviations for the medium and high loads (i.e. 11N and 20N). The test cases 4, 12,

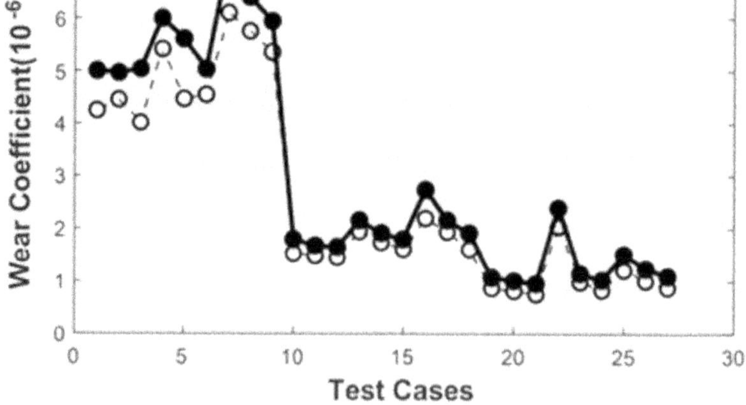

FIGURE 6.4 Graphic representation of wear coefficient analysis of ABS specimens under the horizontal orientation condition of ABS specimens.

TABLE 6.5
Performance Analysis of Human Weight Loss and ANN Weight Loss Analysis in the Vertical Direction

Run	Load (N)	Speed (RPM)	Distance (m)	Wear volume (mm^3)	Wear coefficient ($mm^3/N \cdot m$)
1	2	150	800	5963	3.7270
2	2	150	2150	15783	3.6705
3	2	150	3500	26141	3.7345
4	2	425	800	6522	4.0760
5	2	425	2150	17164	3.9917
6	2	425	3500	28470	4.0472
7	2	700	800	7103	4.4394
8	2	700	2150	19442	4.5189
9	2	700	3500	31805	4.5459
10	11	150	800	29313	4.5768
11	11	150	2150	79348	4.5784
12	11	150	3500	129800	4.7000
13	11	425	800	32587	5.9344
14	11	425	2150	89820	5.7275
15	11	425	3500	145891	5.8227
16	11	700	800	36742	6.6620
17	11	700	2150	98045	6.5677
18	11	700	3500	159406	6.5535
19	20	150	800	43809	2.7381
20	20	150	2150	117996	2.7432
21	20	150	3500	191906	2.7429
22	20	425	800	49759	3.1099
23	20	425	2150	133688	3.1088
24	20	425	3500	216739	3.0980
25	20	700	800	52724	3.2953
26	20	700	2150	142488	3.3057
27	20	700	3500	235056	3.3497

20, 24, and 27 achieved an accuracy of 90%. However, some test cases (i.e. test cases 5 and 6) had a moderate accuracy level in the prediction process. In terms of the overall evaluation, the ANN-centric model obtained higher reliability and precision in predicting vertical weight loss.

Figure 6.5 graphically depicts the experimentally measured wear coefficients and those predicted by the ANN model under the vertical sliding condition of ABS specimens. Superior performance was attained between the ANN model based wear coefficient and the human-based wear coefficient prediction. The ANN model predicted the wear coefficient with higher accuracy, with most values falling within an error margin of 10–15%. A few test cases, like 5, 6, 10, 11, and 12 had the largest

TABLE 6.6
Performance Analysis of Human Wear Coefficient and ANN Wear Coefficient Analysis in the Vertical Direction

Run	Estimated K (mm³/N·m)	ANN predicted K (mm³/N·m)	Error (%)
1	3.7270	3.8005	1.97%
2	3.6705	3.6003	1.91%
3	3.7345	3.7084	0.70%
4	4.0760	4.0931	0.42%
5	3.9917	4.0506	1.47%
6	4.0472	4.1019	1.35%
7	4.4394	4.3542	1.92%
8	4.5189	4.4230	2.12%
9	4.5459	4.4534	2.03%
10	4.5768	4.4897	1.90%
11	4.5784	4.6945	2.54%
12	4.7000	4.7357	0.76%
13	5.9344	5.9012	0.56%
14	5.7275	5.8119	1.47%
15	5.8227	5.7805	0.72%
16	6.6620	6.7503	1.33%
17	6.5677	6.4805	1.33%
18	6.5535	6.6110	0.88%
19	2.7381	2.8016	2.32%
20	2.7432	2.7247	0.68%
21	2.7429	2.7591	0.59%
22	3.1099	3.0790	0.99%
23	3.1088	3.1749	2.13%
24	3.0980	3.1522	1.75%
25	3.2953	3.3740	2.39%
26	3.3057	3.4024	2.93%
27	3.3497	3.4375	2.62%

difference between the human and ANN predictions. As per the analysis, a minor difference was only presented between the ANN-predicted and human-predicted models.

In Table 6.7, the comparative analysis of average wear behavior in both the horizontal and vertical orientations is signified, thereby assessing results from both experimental (human-observed) and ANN-predicted data. In the investigation, higher human wear coefficients (3.10×10^{-6} vs. 2.81×10^{-6}) were exhibited slightly by the vertical orientation of the ABS specimens, thus representing more material degradation under vertical loading. The weight loss of the two orientations of the design was nearly the same. The ANN prediction model was close to the human-centered measurement with respect to accuracy. Centered on the robustness of the model, the

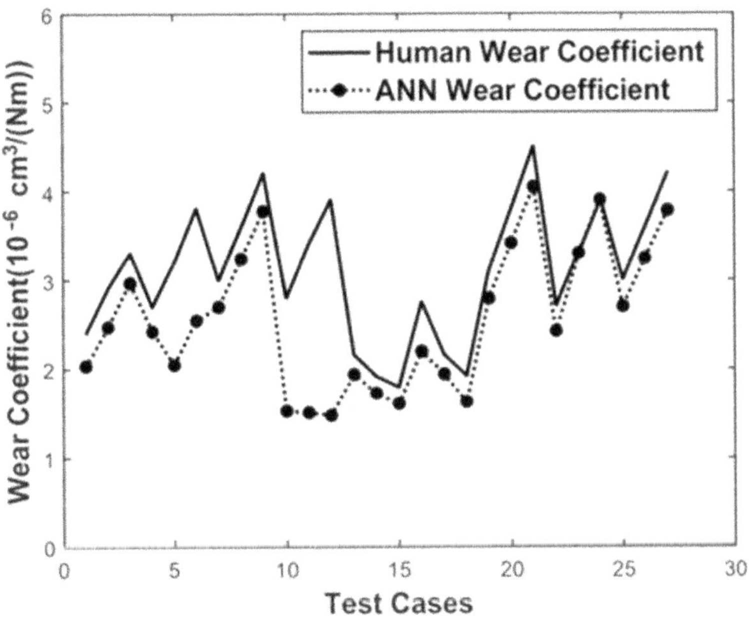

FIGURE 6.5 Pictorial representation of wear coefficient analysis by human-driven mechanism and ANN-based prediction under vertical sliding orientation condition of ABS.

TABLE 6.7
Performance Assessment of Average Wear Coefficient and Weight Loss by Human Experiment and ANN Modeling

Orientation	Average human wear coefficient ($\times 10^{-6}$ cm³/(Nm))	Average human weight loss (mg)	Average ANN wear coefficient ($\times 10^{-6}$ cm³/(Nm))	Average ANN weight loss (mg)
Horizontal	4.6107	100648	4.6036	100254
Vertical	4.3554	95314	4.3571	95108

Where, weight loss (mg) = wear volume (mm^3) x density (g/cm^3) x 1000
Density of ABS =1.04g/cm^3

ANN model outperformed the human-centered model slightly in horizontal orientation with small error margins.

Figure 6.6 displays the coefficient of determination (R^2) analysis for the training, testing, validation, and overall process. All the phases had an R^2 value of above 0.98, thereby signifying that the model predicted more than 98% of the variance in the wear prediction data. According to the testing R^2 value (i.e. 0.981), the ANN model

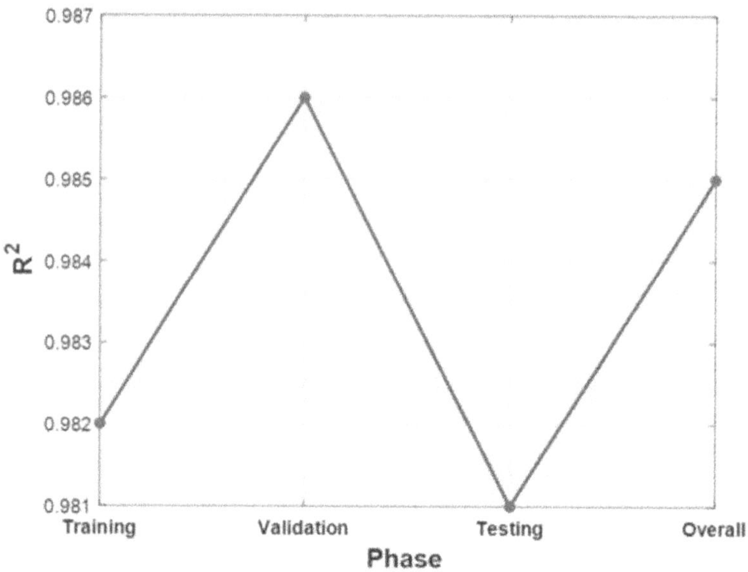

FIGURE 6.6 R^2 analysis.

maintained robust performance even on new data, depicting high reliability and predictive stability. The model consistency and effectiveness of the research were confirmed in the overall analysis phase.

The error histogram analysis of the presented ANN model-based prediction is provided in Figure 6.7. Most of the errors fell between -0.02 and +0.02, thus displaying a minimal deviation and tight error distribution. All error values ranged between -0.03 and +0.03, thereby confirming the existence of very small prediction errors.

The MSE analysis of the (a) horizontal orientation and (b) vertical orientation design of the ABS specimens is shown in Figure 6.8. The MSE value was investigated for training, testing, and validation processes. The MSE performance varied based on the number of epochs. When the epoch count was increased, the MSE value decreased for both orientations of the design. For instance, when the epoch count was 5, the MSE values of the testing, training, and validation were nearly close to the 0.7 value. The investigation displayed that an accurate result was provided by the model prediction.

The regression plot analysis for horizontal orientation and vertical orientation is displayed in Figure 6.9. The training, validation, and testing data were close to the ideal correlation line, representing that an accurate prediction was attained in the investigation. Overall, the regression plot displayed that a strong linear relationship existed between the predicted and measured wear coefficients for both horizontal and vertical orientations of the ABS specimens.

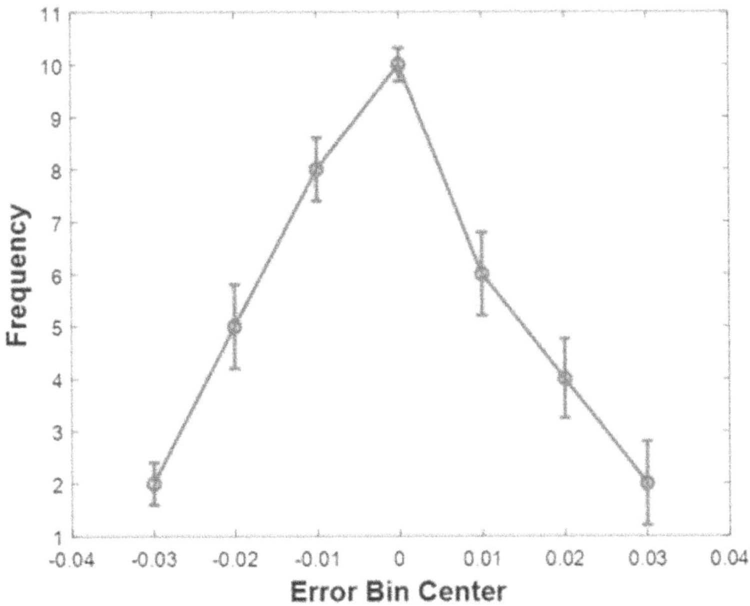

FIGURE 6.7 Error histogram analysis.

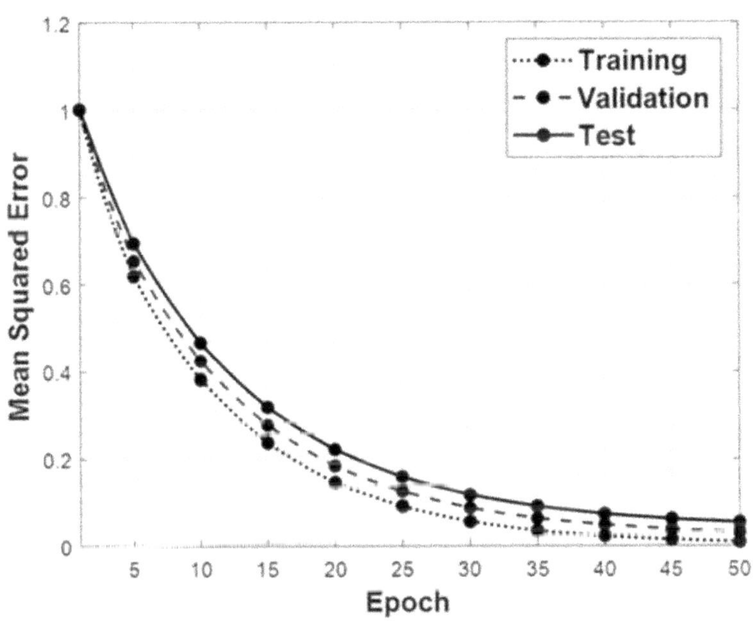

FIGURE 6.8A Mean square error analysis of (a) horizontal orientation design.

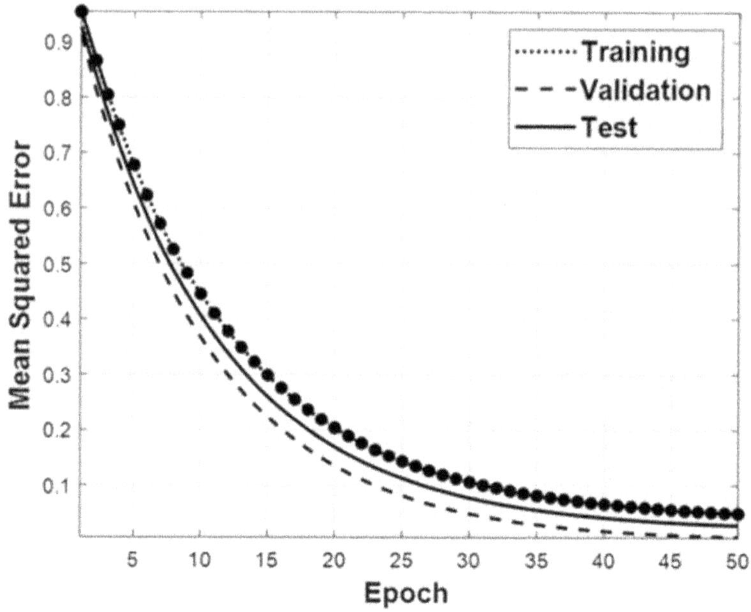

FIGURE 6.8B Mean square error analysis of (b) vertical orientation design.

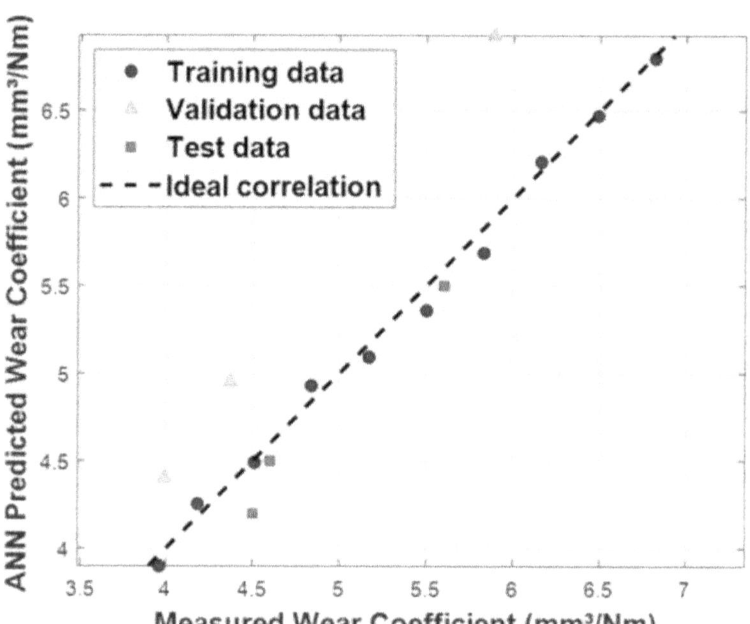

FIGURE 6.9A Regression plot analysis for (a) horizontal orientation.

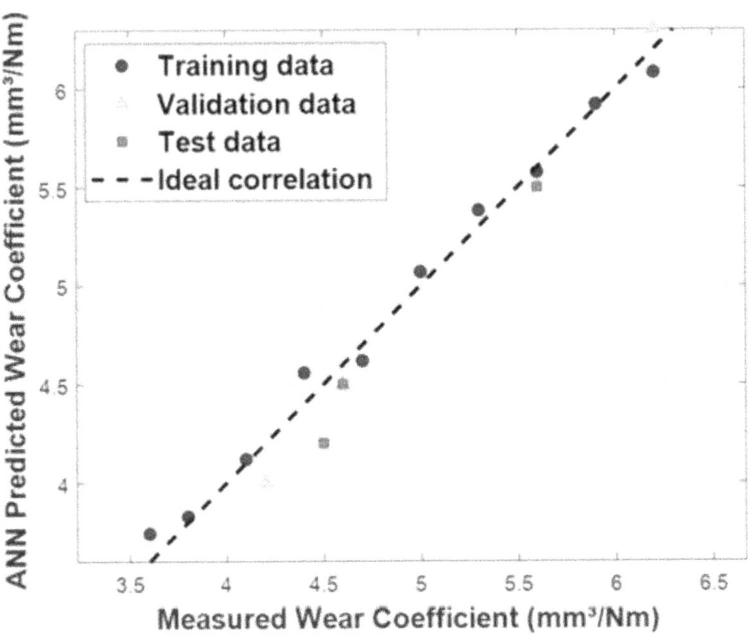

FIGURE 6.9B Regression plot analysis for (b) vertical orientation.

6.5 DISCUSSION

Developing and assessing an ANN model to predict wear behavior under varying load and orientation conditions are the main aims of the presented research. Centered on weight loss, wear coefficient, R^2, and error analysis, the performance of the model was examined against the human-measured value. Across all phases, the ANN model illustrated excellent predictive accuracy. The R^2 values for training (0.982), validation (0.986), testing (0.981), and overall performance (0.985) represented that the model successfully captured the relationship between input parameters (e.g. load, distance, and orientation) and wear behavior. As per these high R^2 values, the model accurately predicted more than 98% of the variance in experimental results. Similarly, the hybrid approaches in ABS obtained the minimum wear rate without considering the different orientations of ABS (Phogat et al., 2022). When analogized with the horizontal orientation (4.6107×10^{-6} cm³/Nm), the vertical orientation displayed a slightly higher average human wear coefficient (4.3554×10^{-6} cm³/Nm), thereby signifying a marginally more aggressive wear mechanism in the vertical configuration. As emphasized in Norani et al. (2021) and Eyri et al. (2025), layer thickness also influenced the wear rate performance. Across all test cases, ANN predictions were consistently close to the experimental values. In the horizontal and vertical orientations, ANN estimations were slightly lower at 4.6036×10^{-6} and 4.3571×10^{-6}, whereas the average human wear coefficients were 4.6107×10^{-6} and 4.3554×10^{-6}, correspondingly. Likewise, average human weight loss (100648-95314 mg) closely matched the outputs of ANN (100254-95108 mg). According to the outcomes, the ANN model could efficiently replicate complicated experimental results with minimal deviation. The studies (Kadauw, 2025; Rasheed et al., 2023) also proved that, for the materials, a better wear rate was rendered by the ANN model and the Taughi optimization process.

6.6 CONCLUSION

In this study, the wear behavior analysis of horizontally and vertically printed ABS specimens was examined utilizing pin-on-disk testing and ANN modeling. Based on the printed ABS specimens in the horizontal and vertical orientations, the wear testing was initially performed using pin-on-disk testing. Afterwards, humans evaluated the wear rate, which was further evaluated using ANN modeling. As per the experimental results, wear characteristics were significantly affected by orientation, with vertically printed specimens depicting slightly higher wear coefficients. This represented increased surface degradation under similar loading and sliding conditions. Moreover, the ANN model effectively captured complex relationships between input parameters (load, sliding distance, and orientation) and wear output (weight loss and wear coefficient) when contrasted with the human analysis. The model depicted excellent accuracy and generalization ability, with R^2 values exceeding 0.98 across training, validation, testing, and overall phases. Furthermore, the error investigation between the measured and ANN prediction revealed that the minimum deviation was presented in the analysis. For the horizontal orientation, the average weight loss and wear coefficient of the ANN model were 100254 mg and 4.6×10^{-6}, correspondingly. According to the overall analysis, the research study assisted in understanding the wear behavior of the material under different conditions.

6.7 FUTURE SCOPE

Nevertheless, the study failed to focus on other factors based on the performance variation of the wear behavior of the material (i.e. geometrical properties and layer thickness). Therefore, the analysis presented will be improved in the future by considering those factors centered on wear behavior analysis.

REFERENCES

Aida, H. J., Nadlene, R., Mastura, M. T., Yusriah, L., Sivakumar, D., & Ilyas, R. A. (2021). Natural fibre filament for fused deposition modelling (FDM): A review. *International Journal of Sustainable Engineering*, *14*(6), 1988–2008. https://doi.org/10.1080/19397038.2021.1962426

Alami, A. H., Ghani Olabi, A., Alashkar, A., Alasad, S., Aljaghoub, H., Rezk, H., & Abdelkareem, M. A. (2023). Additive manufacturing in the aerospace and automotive industries: Recent trends and role in achieving sustainable development goals. *Ain Shams Engineering Journal*, *14*(11), 1–18. https://doi.org/10.1016/j.asej.2023.102516

Albahkali, T., Abdo, H. S., Salah, O., & Fouly, A. (2023). Adaptive Neuro-Fuzzy-based models for predicting the tribological properties of 3D-printed PLA green composites used for biomedical applications. *Polymers*, *15*(14), 1–16. https://doi.org/10.3390/polym15143053

Batista, M., Blanco, D., Del Sol, I., Piñero, D., & Vazquez, J. M. (2021). Tribological characterization of fused deposition modelling parts. *IOP Conference Series: Materials Science and Engineering*, *1193*(1), 1–9. https://doi.org/10.1088/1757-899x/1193/1/012068

Chhabra, D., Deswal, S., Kaushik, A., Garg, R. K., Kovács, A., Khargotra, R., & Singh, T. (2023). Analysis of fused filament fabrication parameters for sliding wear performance of carbon reinforced polyamide composite material fabricated parts using a hybrid heuristic tool. *Polymer Testing*, *118*, 1–11. https://doi.org/10.1016/j.polymertesting.2022.107910

Daminabo, S. C., Goel, S., Grammatikos, S. A., Nezhad, H. Y., & Thakur, V. K. (2020). Fused deposition modeling-based additive manufacturing (3D printing): Techniques for polymer material systems. *Materials Today Chemistry*, *16*, 1–23. https://doi.org/10.1016/j.mtchem.2020.100248

Ergene, B., İnci, Y. E., Çetintaş, B., & Daysal, B. (2025). An experimental study on the wear performance of 3D printed polylactic acid and carbon fiber reinforced polylactic acid parts: Effect of infill rate and water absorption time. *Polymer Composites*, *46*, 372–386. https://doi.org/10.1002/pc.28993

Eyri, B., Gul, O., Yilmaz, S., Karsli, N. G., & Yilmaz, T. (2025). Characterization of ABS/PETG multi-material composites 3D printed by print-pause-print method. *Polymer Engineering and Science*, *65*, 2353–2369. https://doi.org/10.1002/pen.27151

González-Hernández, A., Morales-Cepeda, A. B., Caicedo, J. C., Amaya, C., & Olive-Méndez, S. F. (2022). Structure, functional groups analysis and tribo-mechanical behavior of carbide and nitride coatings deposited on AISI 1060 substrates by RF-magnetron sputtering. *Journal of Materials Research and Technology*, *18*, 5432–5443. https://doi.org/10.1016/j.jmrt.2022.04.075

Guenther, E., Kahlert, M., Vollmer, M., Niendorf, T., & Greiner, C. (2021). Tribological performance of additively manufactured aisi h13 steel in different surface conditions. *Materials*, *14*(4), 1–10. https://doi.org/10.3390/ma14040928

Iftekar, S. F., Aabid, A., Amir, A., & Baig, M. (2023). Advancements and limitations in 3D printing materials and technologies: A critical review. *Polymers*, *15*(11), 1–23. https://doi.org/10.3390/polym15112519

Kadauw, A. A. A. (2025). Neural network optimization of mechanical properties of ABS-like photopolymer utilizing stereolithography (SLA) 3D printing. *Journal of Manufacturing and Materials Processing*, *9*(4), 1–16. https://doi.org/10.3390/jmmp9040116

Keshavamurthy, R., Tambrallimath, V., Patil, S., Rajhi, A. A., Duhduh, A. A., & Khan, T. M. Y. (2023). Mechanical and wear studies of boron nitride-reinforced polymer composites developed via 3D printing technology. *Polymers*, *15*(22), 1–16. https://doi.org/10.3390/polym15224368

Keshavamurthy, R., Tambrallimath, V., Rajhi, A. A., Shabbir Ahmed, R. M., Patil, A. Y., Yunus Khan, T. M., & Makannavar, R. (2021). Influence of solid lubricant addition on friction and wear response of 3d printed polymer composites. *Polymers*, *13*(17), 1–13. https://doi.org/10.3390/polym13172905

Kumar, A., Choudhary, A., Tiwari, A., James, C., Kumar, H., Kumar Arora, P., & Akhtar Khan, S. (2021). An investigation on wear characteristics of additive manufacturing materials. *Materials Today: Proceedings*, *47*, 3654–3660. https://doi.org/10.1016/j.matpr.2021.01.263

Mahmood, A., Akram, T., Chen, H., & Chen, S. (2022). On the evolution of additive manufacturing (3D/4D printing) technologies: materials, applications, and challenges. *Polymers*, *14*(21), 1–31. https://doi.org/10.3390/polym14214698

Nataraja, M., Balakumar, G., Santhosh, N., & Rudra Naik, M. (2024). Characterization of wear rate of Al-12 wt%Si alloy based MMC reinforced with ZrO2 particulates. *Materials Research Express*, *11*(3), 1–21. https://doi.org/10.1088/2053-1591/ad3468

Njim, E. K., Jweeg, M. J., Al-Maamori, M. H., Idan, Z. S., Al-Waily, M., Mouthanna, A., & Hadji, L. (2024). Wear behavior analysis of imperfect functionally graded parts: Analytical and experimental techniques. *BIO Web of Conferences*, *97*, 1–12. https://doi.org/10.1051/bioconf/20249700001

Norani, M. N. M., Abdollah, M. F. B., Abdullah, M. I. H. C., Amiruddin, H., Ramli, F. R., & Tamaldin, N. (2021). 3D printing parameters of acrylonitrile butadiene styrene polymer for friction and wear analysis using response surface methodology. *Proceedings of the Institution of Mechanical Engineers, Part J: Journal of Engineering Tribology*, *235*(2), 1–10. https://doi.org/10.1177/1350650120925601

Pant, M., Singari, R. M., Arora, P. K., Moona, G., & Kumar, H. (2020). Wear assessment of 3-D printed parts of PLA (polylactic acid) using Taguchi design and Artificial Neural Network (ANN) technique. *Materials Research Express*, *7*(11), 1–16. https://doi.org/10.1088/2053-1591/abc8bd

Phogat, A., Chhabra, D., Sindhu, V., & Ahlawat, A. (2022). Analysis of wear assessment of FDM printed specimens with PLA, multi-material and ABS via hybrid algorithms. *Materials Today: Proceedings*, *62*, 37–43. https://doi.org/10.1016/j.matpr.2022.01.429

Portoacă, A. I., Ripeanu, R. G., Diniță, A., & Tănase, M. (2023). Optimization of 3D printing parameters for enhanced surface quality and wear resistance. *Polymers*, *15*(16), 1–22. https://doi.org/10.3390/polym15163419

Rasheed, A., Hussain, M., Ullah, S., Ahmad, Z., Kakakhail, H., Riaz, A. A., Khan, I., Ahmad, S., Akram, W., Eldin, S. M., & Khan, I. (2023). Experimental investigation and Taguchi optimization of FDM process parameters for the enhancement of tensile properties of Bi-layered printed PLA-ABS. *Materials Research Express*, *10*(9), 1–19. https://doi.org/10.1088/2053-1591/acf1e7

Sapkota, A., Ghimire, S. K., & Adanur, S. (2024). A review on fused deposition modeling (FDM)-based additive manufacturing (AM) methods, materials and applications for flexible fabric structures. *Journal of Industrial Textiles*, *54*, 1–51. https://doi.org/10.1177/15280837241282110

Shah, R., Pai, N., Rosenkranz, A., Shirvani, K., & Marian, M. (2022). Tribological behavior of additively manufactured metal components. *Journal of Manufacturing and Materials Processing*, *6*(6), 1–15. https://doi.org/10.3390/jmmp6060138

Singh, M., & Bharti, P. S. (2021). Parametric influence of process parameters on the wear rate of 3D printed polylactic acid specimens. *Indian Journal of Pure and Applied Physics*, *59*(3), 244–251.

Slapnik, J., Stiller, T., Wilhelm, T., & Hausberger, A. (2020). Influence of solid lubricants on the tribological performance of photocurable resins for vat photopolymerization. *Lubricants*, *8*(12), 1–19. https://doi.org/10.3390/lubricants8120104

Sola, A., Rosa, R., & Ferrari, A. M. (2024). Environmental impact of fused filament fabrication: What is known from life cycle assessment? *Polymers*, *16*(14), 1–26. https://doi.org/10.3390/polym16141986

Srinivasan, R., Rathish, R., Sivaraman, P. R., Adwaith Pramod, G., & Shivaganesh. (2020). Influential analysis of fused deposition modeling process parameters on the wear behaviour of ABS parts. *Materials Today: Proceedings*, *27*, 1869–1876. https://doi.org/10.1016/j.matpr.2020.03.808

Zakaulla, M., Pasha, Y., & Siddalingappa, S. K. (2021). Prediction of mechanical properties for polyetheretherketone composite reinforced with graphene and titanium powder using artificial neural network. *Materials Today: Proceedings*, *49*, 1268–1274. https://doi.org/10.1016/j.matpr.2021.06.365

7 Surface Engineering of Additive Manufactured Parts Based on Polymer, Metal, and Ceramic Material

Amrinder Singh Johal, Areeb Shazli, Navdeep Sharma Dugala, Dinesh Kumar, and Ashish Kumar

7.1 INTRODUCTION

Additive manufacturing (AM), now recognized as a transformative force in modern production, had its origins in the early 1980s. One of the earliest breakthroughs came from Dr. Hideo Kodama, who proposed a method for rapid prototyping using photopolymer-based layer-by-layer fabrication. This concept lay the foundation for what would become a rapidly advancing field. The journey accelerated in 1984 when Charles Hull introduced stereolithography (SLA), a technique that used ultraviolet (UV) light to solidify liquid resins into 3D structures—marking one of the first practical implementations of AM. The basic workflows of AM and quick prototyping are shown in Figure 7.1 and Figure 7.2, respectively.

What began as a tool for prototyping has since matured into a full-fledged manufacturing solution, powered by continuous innovation in material development, computer-aided design (CAD), and precision control systems. Today, AM is no longer limited to simple models or design testing—it plays an active role in producing high-performance components across the aerospace, automotive, medical, and consumer sectors (Rajaguru et al., 2019).

For polymers, popular techniques like fused deposition modeling (FDM) melt and extrude thermoplastic filaments, offering a low-cost, accessible option for functional prototyping. SLA and digital light processing (DLP) use photopolymerization to deliver high precision and smooth surface finishes, while selective laser sintering (SLS) fuses powdered polymers to form mechanically stable parts. These approaches are well suited for industries where speed, customization, and visual accuracy are critical such as consumer products and automotive design (Abdulhameed et al., 2019; Bikas et al., 2015; Prakash et al., 2018; Rivera & Arciniegas, 2020; Zhang et al., 2018).

DOI: 10.1201/9781003643531-7

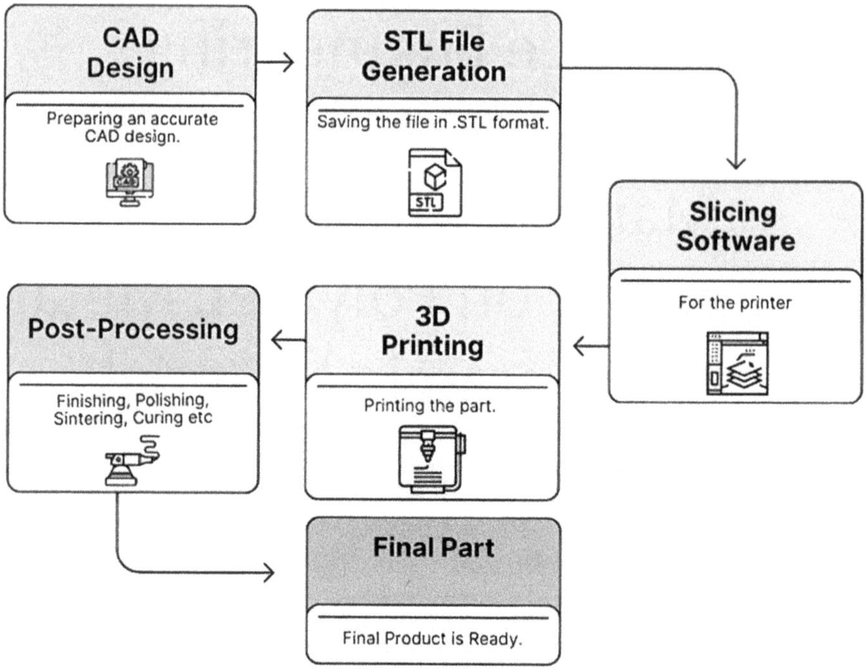

FIGURE 7.1 Basic workflow of additive manufacturing.

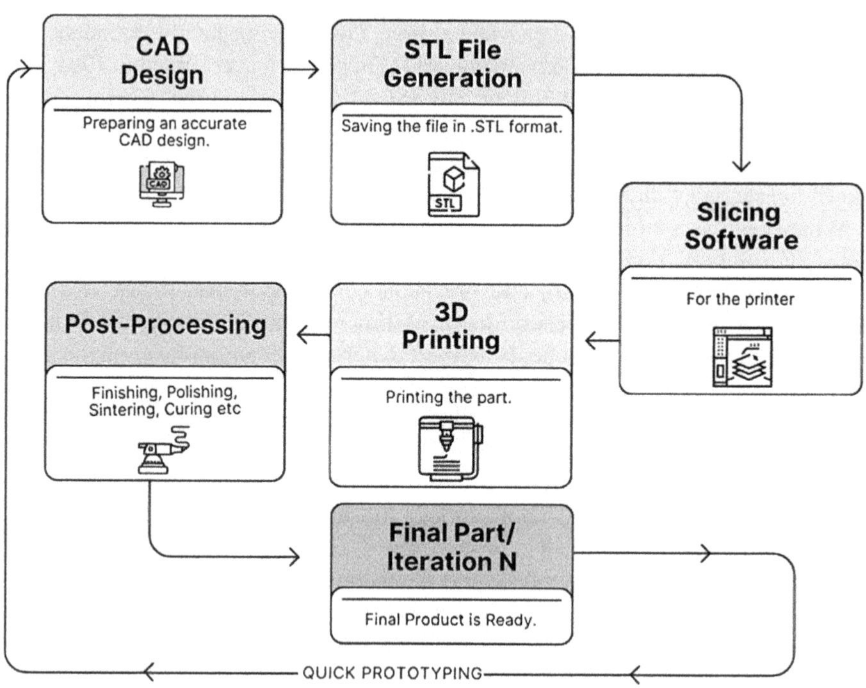

FIGURE 7.2 Representation of quick prototyping.

Metals, by contrast, demand energy-intensive methods due to their high melting points and strength requirements. Powder bed fusion (PBF) techniques like selective laser melting (SLM) and electron beam melting (EBM) rely on concentrated energy sources to fuse fine metal powders into solid, load-bearing components. Direct energy deposition adds another dimension—allowing metallic powder or wire to be deposited onto existing parts for repairs or hybrid builds. These methods are crucial where reliability under stress is non-negotiable, especially in aerospace applications and for medical implants (Abdulhameed et al., 2019; Bikas et al., 2015; Prakash et al., 2018; Rivera & Arciniegas, 2020; Zhang et al., 2018).

Additive manufacturing with ceramics presents a different set of challenges. Due to their inherent brittleness and high sintering temperatures, ceramics require carefully controlled workflows. Techniques such as vat photopolymerization (VPP) using ceramic suspensions, indirect laser sintering (ILS), and binder jetting (BJ) enable initial part formation, followed by sintering to achieve full strength. These technologies are particularly valuable in fields like electronics, biomedicine, and energy systems, where properties like biocompatibility and thermal resistance are essential.

Every material category brings its own set of limitations and trade-offs. While polymers offer design flexibility and affordability, they often fall short in load-bearing or high-temperature conditions. Metal AM, although capable of producing strong, complex parts, introduces concerns such as internal stresses, porosity, and the need for precise support strategies. Ceramic AM workflows are complicated by post-processing steps and material shrinkage during sintering, which demands dimensional compensation during design.

This chapter mainly focuses on the defects that may form when these AM techniques are used. To solve these defects, the AM part needs to be surface engineered, ironing out imperfections and defects on the surface. Some of these defects may include balling, porosity, and roughness. Defects and challenges in AM surface engineering are mentioned in this chapter.

7.2 IMPORTANCE OF SURFACE QUALITY IN AM

Surface quality is a decisive determinant of functional performance in AM parts across industries because it directly influences the fatigue life, wear, corrosion, friction, sealing, biocompatibility, and dimensional conformance of critical features such as channels and interfaces. Unlike conventionally machined parts, AM components often emerge with process-intrinsic surface conditions, partially fused particles, staircase terracing, balling deposits, and surface-connected pores that serve as stress concentrators, initiation sites for cracks, and nuclei for corrosive attack or bacterial colonization, unless engineered by post- or in-process treatments that are tailored to application requirements. These features drive the observed disparity between as-built and finished AM properties, and motivate for robust surface engineering to bridge performance gaps (Brennan et al., 2021; Lee et al., 2020; Ramezani et al., 2023).

In fatigue-critical service, surface integrity modulates crack nucleation, short-crack growth, and endurance limit. Studies show that AM parts can exhibit up to ~60% lower fatigue life than conventionally finished counterparts when as-built

TABLE 7.1
Type of Defects for Additive Manufactured Parts

Sr. No.	Defect type	Root cause	Effect on surface/part	Main processes affected
1.	Porosity	Gas entrapment Incomplete fusion	Low fatigue life; roughness	PBF, DED, BJ
2.	Cracking	Stress Cooling Shrinkage	Fracture; poor durability	PBF, DED
3.	Distortion	Residual stress Cooling	Warping, loss of tolerances	PBF, DED, FDM
4.	Balling	Melt pool instability	Surface roughness, porosity	PBF
5.	Surface roughness	Layering Particle adhesion	High Ra, poor aesthetics	All AM
6.	Delamination	Poor layer bonding	Weakness, part separation	FDM, SLA, some PBF
7.	Residual stress	Thermal cycling Cooling	Cracks, distortion	PBF, DED, SLA

roughness and porosity remain, with surface-connected porosity and lack-of-fusion defects accelerating early crack initiation under cyclic loading. For example, PBF components with Ra in the 9–50 μm range (material- and system-dependent) require post-processing to achieve the fatigue-relevant finishes (e.g. Ra < 1 μm) commonly demanded in aerospace and power train services. The relationship is non-linear: small reductions in peak-to-valley height (Rz) or the removal of particle agglomerates on the edges can deliver disproportionate fatigue benefits by suppressing local notch effects and eliminating surface-breaking pores that bypass the beneficial effect of hot isostatic pressing (HIP) on internal porosity (Brennan et al., 2021; Lee et al., 2020). Different types of defects for AM parts are shown in Table 7.1.

In tribological contexts, the as-built surface controls friction, lubrication regimes, and wear mechanisms. Elevated roughness promotes abrasive and adhesive wear, increases running-in time, and reduces film formation in lubricated contacts, leading to accelerated material loss and temperature rise. Mechanical finishing (machining, grinding, and vibratory/tumbling) and surface peening are widely used to remove asperities, compress near-surface defect populations, and tailor compressive residual stress profiles that retard crack initiation, often cutting Ra by an order of magnitude, and improving wear and scuffing resistance in steels, titanium, and nickel alloys. In energy devices (e.g. fuel cells and heat exchangers), the surface state of channels governs pressure drop, reactant transport, and fouling; smoothing internal passages by abrasive flow machining (AFM) or chemical finishing reduces hydraulic losses and mitigates particulate adhesion, directly improving device efficiency and service life (Lee et al., 2020; Ramezani et al., 2023).

Corrosion behavior is likewise surface-sensitive. Surface-connected pores and roughness amplify the real exposed area, trap electrolytes, and create occluded regions that destabilize passive films, especially in chloride-bearing environments. Smoother, denser surfaces (via laser remelting, electropolishing, or coatings) reduce pit nucleation and stabilize passive layers, improving corrosion current density and critical pitting potential. Protective coatings applied by thermal spray, physical vapor deposition (PVD) or chemical vapor deposition (CVD), or laser cladding add additional barriers, like engineered chemistries (e.g. chromium nitride (CrN), diamond-like carbon (DLC), and ceramic oxides), and tailored microstructures to resist corrosive attack and galvanic interactions in mixed-metal assemblies (Ramezani et al., 2023).

Biomedical implants offer clear case studies. For orthopedics and dental applications, two seemingly contradictory surface demands coexist: macro-porous textures that promote osseointegration, and micro/nano-smoothness that avoid wear debris in articulating regions and suppress bacterial adhesion on soft tissue interfaces. Bioactive and antimicrobial coatings (e.g. hydroxyapatite, Ag- or Zn-containing nanostructures) and texturing can reconcile these requirements, and enable bone ingrowth, while reducing the risk of infection and improving the wear survivability of bearing surfaces. Surface finish also dictates protein adsorption and cell response; engineered topographies and chemistries can bias osteoblast adhesion and proliferation, improve hemocompatibility, and reduce inflammation at the implant-tissue interface (Ramezani et al., 2023).

In aerospace, as-built surfaces influence aerodynamic drag, heat transfer, and structural integrity. Rough external skins increase boundary layer turbulence and fuel burn; roughness-induced hot spots on turbine hot-section parts impair thermal barrier coating performance and accelerate oxidation. Internally, rough cooling passages elevate pressure drops and reduce cooling effectiveness. Achieving Ra < 1 μm on flow surfaces, and < 20 μm Rz on gas-path surfaces is a typical requirement post-AM to meet engine efficiency and durability goals. Surface peening and laser shock peening are further leveraged to instill deep compressive stress profiles that enhance foreign-object damage tolerance and dwell fatigue resistance in titanium alloys (Lee et al., 2020; Ramezani et al., 2023).

Automotive exemplars include fuel injector nozzles and power train gears. Nozzle surface finish controls spray atomization and coking; micro-polishing and electropolishing suppress deposit formation, sustaining emissions compliance and fuel economy. Gear tooth flanks demand low roughness for micro-pitting resistance and noise control; applying PVD hard coatings (e.g. CrN) and finishing grinds produces stable tribofilms and improves scuffing resistance under boundary lubrication. In AM manifolds and casings, finishing and sealing (e.g. impregnations and coatings) are critical to achieve leak-tightness when surface-connected porosity is present (Brennan et al., 2021; Lee et al., 2020; Ramezani et al., 2023).

Finally, geometric fidelity couples to the surface. Stair-step terracing on shallow angles and partially fused particles bias dimensional error and clearance/jamming risks in assemblies. Precision surfaces are realized by hybrid finishing (milling-polishing-electrochemical steps) guided by AM-specific design-for-finish principles (part orientation, feature accessibility, and support layout) and surface metrology feedback (De Pastre et al., 2022; Lee et al., 2020).

The net effect is that surface engineering transforms AM's layer-defined, particle-decorated as-builts into application-grade components. It elevates fatigue, wear, and corrosion performance; enables regulatory acceptance in medical and aerospace industries; and unlocks the full benefits of AM geometry by delivering the required functional interface between the part and the environment.

7.3 SURFACE CHALLENGES IN ADDITIVE MANUFACTURING

Roughness is the most visible challenge across AM modalities. In laser or electron PBF (SLM or direct metal laser sintering (DMLS) and EBM), typical as-built surface roughness spans Ra ≈ 8–50 μm (system- and material-dependent). This is driven by partially sintered particles, staircase effects, spatters, and melt pool instabilities (balling), with EBM often exhibiting coarser textures than laser PBF. Surface roughness on the upskin or sidewalls differs from downskin regions, supported by unfused powder; downskin regions tend to trap partially fused particles and sag due to reduced heat dissipation and poor overhang stability. In DED (laser or electron), larger melt pools and higher deposition rates yield coarser bead ripples and heightened waviness, particularly at inter-bead boundaries; lower deposition rates and controlled overlap can reduce macro-roughness, but at a cost to productivity. In FDM or fused filament fabrication (FFF), the bead profile and layer height dictate the staircase amplitude; polymer stringing and filament under- or over-extrusion cause surface non-uniformities (De Pastre et al., 2022; Lee et al., 2020).

Porosity appears in several forms (Figure 7.3):

- Internal gas porosity from entrapped shielding/powder gases
- Lack-of-fusion (LOF) defects—irregular, elongated voids between scan tracks or layers—due to insufficient energy density or excessive hatch spacing
- Keyhole pores formed by beam-induced vapor cavities that collapse and are captured during rapid solidification.

Surface-connected porosity uniquely affects surfaces: pores intersecting the free surface are not eliminated by HIP and become crack initiators, leak paths, or corrosion

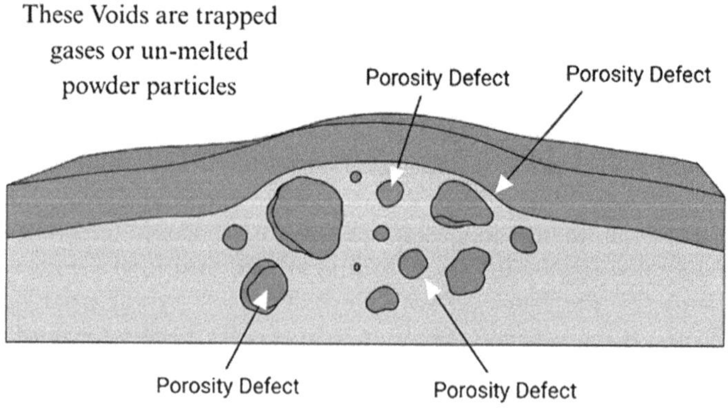

FIGURE 7.3 Porosity defect.

pits. In PBF, process maps (power, speed, hatch, and layer thickness) balance full melting against keyholing; in DED, powder catch efficiency, travel speed, and beam focus influence melt pool wetting and the remelting of prior beads; in FDM, lack of bead fusion and inter-road voids produce surface porosity when temperature and raster overlap are inadequate (Brennan et al., 2021; De Pastre et al., 2022).

Residual stress arises from steep thermal gradients inherent to localized energy input and layer-wise reheating, leading to distortion (warping on supports) (Figure 7.4), cracking (solidification or solid-state) (Figure 7.5), and premature failure if not managed. The thermally graded mechanism in PBF induces tensile stress upon cooling of the top layers, constrained by the cooler bulk; scanning strategy, preheat, layer thickness, and baseplate restraint modulate magnitudes and distributions.

Solidification cracking, particularly in alloys with wide freezing ranges or susceptible compositions (e.g. Nb-bearing superalloys), initiates along inter-dendritic liquid films under shrinkage strain; susceptibility depends on the brittle temperature range and the ability for liquid backfilling. Ductility dip cracking and heat-affected cracking

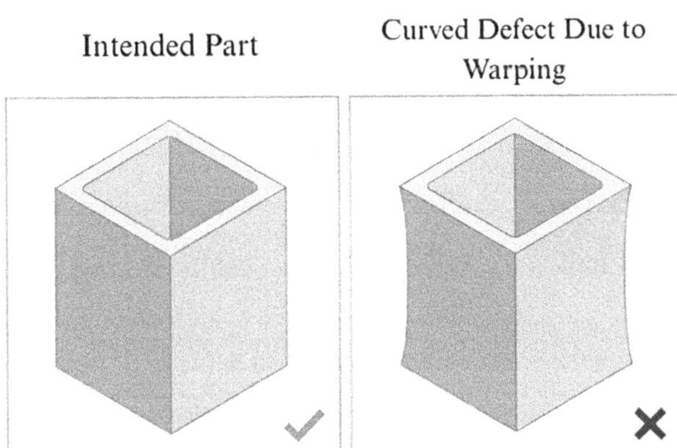

FIGURE 7.4 Distortion and warping.

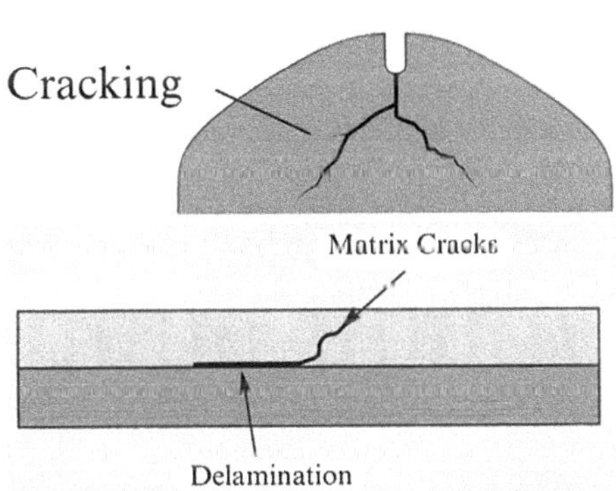

FIGURE 7.5 Cracking defect (hot/cold/delamination).

can arise during post-weld heat treatment or in-process reheating for certain nickel alloys (Chen et al., 2021).

Microstructural defects include epitaxial columnar grains with strong textures, which can introduce anisotropy, the segregation of alloying elements to inter-dendritic regions, oxide inclusions, and unmelted or partially melted particles at surfaces. Balling (Rayleigh instability of the melt track) generates discontinuous beads and spherical droplets that solidify on the surface, increasing roughness and creating near-surface defects. Powder morphology (angular vs. spherical), particle size distribution, and layer packing affect surface quality: spherical, narrowly distributed powders flow and pack more uniformly, reducing surface nodules and improving downskin quality; large particles or satellites increase asperity formation. Process gas flow and spatter dynamics also contaminate sidewalls, depositing resolidified droplets on surfaces (Brennan et al., 2021; Lee et al., 2020; Rasiya et al., 2021). Surface integrity challenges across AM processes are shown in Table 7.2.

Origins are fundamentally tied to localized melting/solidification or deposition and the powder-bed or filament topology. The free surface of the melt pool interacts with underlying roughness, modifying capillary-driven flow and wetting; remelting/proximity heating affects particle adhesion at the sidewalls; and the discontinuous layer-wise nature yields terraces on sloped features. These causes explain why AM requires a surface engineering plan (Figure 7.6) that is tuned to the process, alloy, and geometry to close performance gaps (De Pastre et al., 2022; Lee et al., 2020).

TABLE 7.2
Surface Integrity Challenges Across Additive Manufacturing Processes

Process	Typical surface roughness (Ra)	Key drivers and defects	Microstructural features and stresses
Laser PBF (SLM/DMLS)	≈ 8–17 μm (varies); up to ~50 μm	Partially sintered particles, balling, spatter, **staircase**; **porosity** (gas, LOF, keyhole)	Strong **residual stress**; anisotropic columnar grains
EBM	19–30 μm (horizontal), 24–39 μm (vertical)	Higher layer thickness and preheat sintering; particle adherence	Typically lower residual stresses; coarser microstructure
DED (Laser/e-beam)	Characterized by bead ripples	Bead ripples, powder feeding issues, gas entrapment; larger LOF	Significant **residual stress**; oxide inclusions
FDM/FFF (polymers)	Dominated by layer height and bead geometry	Deformed edges, under-extrusions, weak fiber bonding, base bulging; **staircase** on shallow angles	Anisotropic bonding; roughness variability
Binder jetting (metals/ceramics)	Relatively fine, but fragile as-printed	Sintering-induced shrinkage and grain growth; unbound/loosely bonded particles	Surface finish changes after sintering; infiltration often required

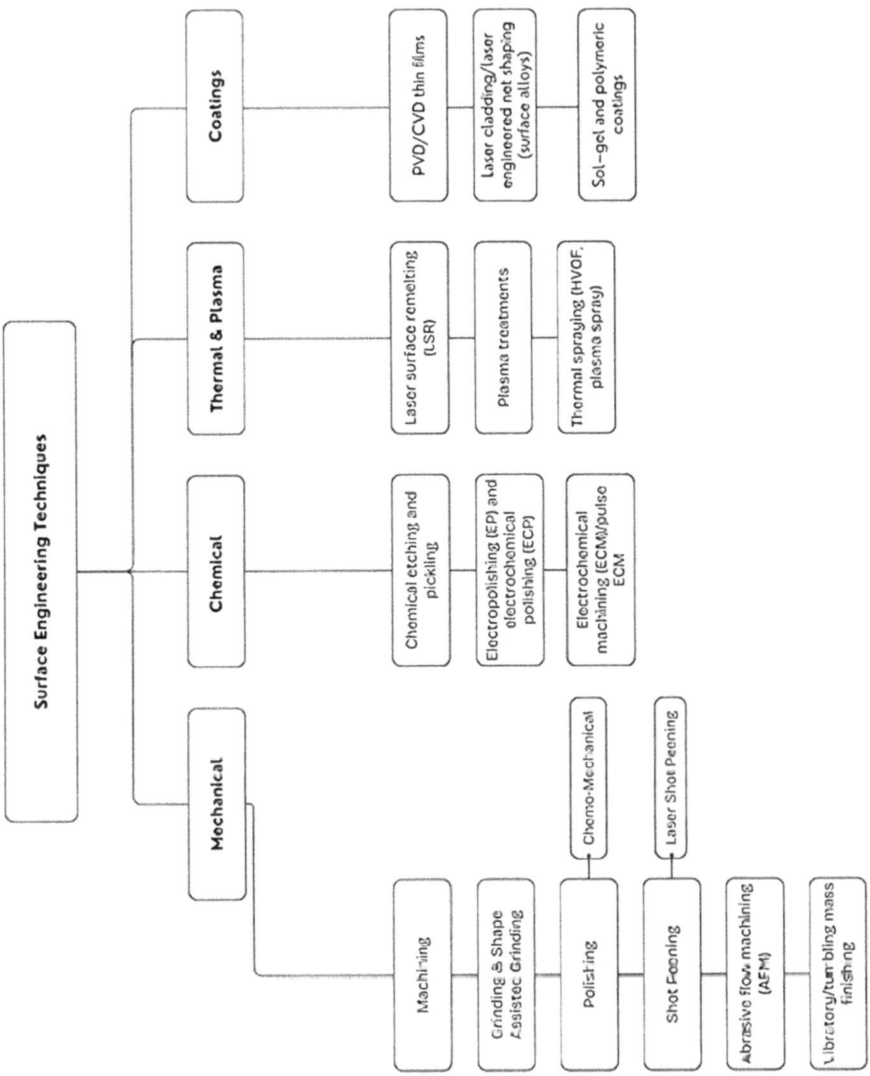

FIGURE 7.6 Surface engineering techniques.

7.4 MECHANICAL POST-PROCESSING

7.4.1 Machining

The principle of machining entails material removal via cutting to achieve dimensional accuracy and a smooth surface finish. In AM, machining is frequently applied to sealing faces, interfaces, and datum features. Parameters include tool geometry/material, cutting speed, feed, depth of cut, coolant strategy, and fixture stiffness. Benefits include deterministic tolerances, low Ra (<1 μm achievable), and stable process knowledge. Disadvantages include access limitations for internal features, potential for burrs and tensile surface stress unless controlled, and the risk of delamination in porous or thin-walled regions. AM adaptations include design-for-machining (stock allowance and probe targets), hybrid AM computer numerical control (CNC) cells for in-chuck transitions, and post-build stress relief prior to finishing to stabilize geometry (Lee et al., 2020).

7.4.2 Grinding and Shape-adaptive Grinding (SAG)

The principle of grinding and shape-adaptive grinding (SAG) entails abrasive removal with bonded wheels or compliant tooling. SAG uses a soft tool backed by controlled pressure to conform to complex geometries, distributing abrasives across the surface. Benefits include sub-micron finishes (e.g. down to <10 nm Ra reported on Ti-6Al-4V plates), suitability for hardened surfaces and complex curvatures, and consistent waviness reduction. Disadvantages include setup complexity, potential thermal damage without cooling, and it being slow for large stock removal. AM adaptations include computer-aided systems' (CAS)-driven toolpath generation from STL/scan data, robotic multi-axis systems to address free forms, and staged abrasives to transition from Ra ~10–50 μm to Ra < 1 μm (Lee et al., 2020).

7.4.3 Polishing (Mechanical and Chemo-mechanical)

The principle of polishing (mechanical and chemo-mechanical) entails fine abrasive slurries or pads to remove asperities and produce mirror finishes; chemo-mechanical variants rely on synergistic chemical softening plus mechanical action. Benefits include best-in-class finish quality for optics/biomedical devices, and low induced subsurface damage. Disadvantages include surface-only access, consumable management, the potential rounding of edges, and the loss of sharp detail. AM adaptations include localized pad tools for internal features, which can be reached via ports, near-net AM to reduce polishing time, and integration with metrology to avoid over-polishing (Lee et al., 2020).

7.4.4 Shot Peening/Laser Shock Peening

The principle of shot peening or laser shock peening entails the impact loading of the surface (mechanical or laser-induced plasma) to cold work and impart compressive residual stress, refine near-surface microstructure, and close microcracks. Parameters include coverage, intensity (Almen arc height), media size/hardness for shot peen, pulse energy, spot size, overlap, and confinement water layer for laser peen. Benefits include improved fatigue strength, and foreign object damage (FOD) tolerance, which is beneficial in AM to counter tensile residual stresses and notch sensitivity. Disadvantages include surface roughening for shot peen, process control, which is required to avoid damage, and access limitations. AM adaptations include sequence peening after roughness removal to maximize fatigue benefits, and laser peening for high-value aero/energy parts with complex shapes (Lee et al., 2020; Ramezani et al., 2023).

7.4.5 Abrasive Flow Machining/Magnetic-assisted Variants

The principle of abrasive flow machining entails viscoplastic media with embedded abrasives extruded through internal channels to smooth interiors. Magnetic-assisted AFM (MAAFM) or magnetorheological AFM (MRAFF) use magnetic fields to stiffen or shape the media or control cutting forces. Parameters include pressure, media viscosity, abrasive type/size, and the number of cycles. Benefits include the unique ability to finish internal passages, a reduction of Ra from ~10–30 μm to <1 μm, demonstrated on Ti64 channels, and improved flow efficiency. Disadvantages include tool access (inlet/outlet), potential damage to thin walls at high pressure, and media cleanup. AM adaptations include the design of auxiliary ports for media flow, and staged AFM, followed by chemical passivation for corrosion-sensitive channels.

7.4.6 Vibratory/Tumbling Mass Finishing

The principle of vibratory/tumbling mass finishing entails bulk abrasive action in vibratory bowls or barrel finishing to smooth external surfaces, and peening-polish hybrids (tribofinishing). Benefits include it being cost-effective for batches, as well as substantial roughness reductions (e.g. ~70% improvement reported) and fatigue benefits. Disadvantages include edge rounding, limited control for precision features, large media consumption and long cycle times (up to 48 hours) for hard alloys. AM adaptations include pre-machining critical faces, masking features that require dimensional control, and combining with peening for fatigue.

7.5 CHEMICAL POST-PROCESSING

7.5.1 Chemical Etching and Pickling

The principle of chemical etching and pickling entails selective chemical dissolution to remove surface-adherent particles, recast layers or contamination, and acids/alkalis tailored to the alloy. Benefits include uniform treatment on complex geometries, and the removal of loosely bonded features. Disadvantages include limited finish improvement (etch pits), material loss, and waste management. AM adaptations include pre-etching to debond particles before mechanical finishing, and use for fine lattice cleaning (Ramezani et al., 2023).

7.5.2 Electropolishing and Electrochemical Polishing

The principle of electropolishing (EP) and electrochemical polishing (ECP) entails anodic dissolution under mass transport-limited conditions, and leveling by preferential removal at asperity peaks. The parameters include electrolyte composition (e.g. acids for Ti, Ni, SS), temperature, current density, voltage, time, and agitation. Benefits include significant Ra reduction, improved corrosion resistance via passivation, low mechanical damage, and it being suitable for internal line-of-sight limited regions with electrode design. Disadvantages include alloy-specific chemistries, hydrogen pickup risk for some alloys, and geometry-dependent current distribution. AM adaptations include tailored electrolytes for AM microstructures, pulsed EP to reduce edge effects, combination with passivation steps, and reported finishes to Sa < 25 nm on DMLS SS316L after pre-machining plus fine abrasives.

7.5.3 Electrochemical Machining/Pulse Electrochemical Machining

The principle of electrochemical machining (ECM)/pulse ECM entails anodic dissolution with shaped cathodes to form or finish features; pulsed ECM improves control. Benefits include no tool forces, excellent finishes on hard alloys, and minimal recast. Disadvantages include tooling complexity, electrolyte handling, and its accuracy being dependent on gap control. AM adaptations include the die-sinking of ports, and the smoothing of inaccessible surfaces in conjunction with AM near-net geometry (Ramezani et al., 2023).

7.6 THERMAL AND PLASMA-BASED METHODS

7.6.1 Laser Surface Remelting

The principle of laser surface remelting (LSR) entails the localized laser melting of the top layer to promote surface tension-driven leveling, dissolve near-surface pores, and refine microstructure. Parameters include power, speed, overlap, beam size, and atmosphere. Benefits include that it reduces roughness, seals surface-connected porosity, homogenizes chemistry, and can be integrated in-process. Disadvantages include the risk of keyholing or microcracking if parameters are mis-set, the heat-affected zone, and limited reach in deep channels. AM adaptations include that the *in-situ* skin remelt passes on sidewalls or downskins during PBF, and post-process contour remelt on DED build-up for bead leveling (Dahotre et al., 2016; De Pastre et al., 2022).

7.6.2 Plasma Treatments

The principle of plasma treatments entails ionized gas exposure to clean, activate, or deposit thin functional layers; it includes plasma nitriding or carburizing for diffusion hardening, and plasma electrolytic treatments. Benefits include that it improves hardness, wear, and fatigue, and enhances adhesion for coatings. Disadvantages include thermal exposure, geometry shielding, and process controls. AM adaptations include low-temperature plasma nitriding for stainless steel/titanium to maintain bulk properties, while enhancing the surface, and plasma electrolytic polishing for conductive alloys to smooth complex shapes (Ramezani et al., 2023).

7.6.3 Thermal Spraying (High-velocity Oxygen Fuel, Plasma Spray)

The principle of thermal spraying entails accelerated molten/semi-molten particles deposited to form coatings (metals, ceramics, cements). Parameters include feedstock, particle velocity/temperature, stand-off distance, and substrate preparation. Benefits include thick, protective coatings (wear, corrosion, and thermal barriers), high deposition rates, and it being applicable to AM substrates. Disadvantages include porosity and residual stress, it requiring grit-blasting (managed for AM), and adhesion being critical. AM adaptations include bond coat + top-coat stacks (e.g. MCrAlY + yttria-stabilized zirconia (YSZ)) on AM hot-section parts, and peening or ultrasonic nanocrystalline surface modification pre-spray to improve adhesion on tool steels (Dahotre et al., 2016).

7.7 COATING TECHNOLOGIES (METALLIC, CERAMIC, POLYMER COATINGS, THIN FILMS, SURFACE ALLOYS)

7.7.1 PVD/CVD Thin Films

The principle of PVD/CVD thin films entails the vapor-based deposition of hard, dense films (e.g. dense titanium nitride (TiN), CrN, DLC) via physical (sputter/evaporation) or chemical routes. Benefits include low roughness addition, high hardness, corrosion resistance, and it being good for wear and tribology. Disadvantages include line-of-sight limitations for PVD, and high temperatures for some CVD; adhesion depends on substrate preparation. AM adaptations include micro-blasting plus PVD for gears/valves from AM, multilayer architectures to accommodate AM substrate roughness, and conformal CVD for internal features where feasible (Ramezani et al., 2023).

7.7.2 Laser Cladding/Laser-engineered Net Shaping (Surface Alloys)

The principle of laser cladding or laser-engineered net shaping entails concurrent powder/wire feed and laser melting to deposit dense metallurgically bonded overlays or composites; in surface engineering and advanced materials (SEAM), the *in-situ* synthesis of alloys or carbides is possible. Benefits include site specific overlays (e.g. stellite, Inconel, vanadium carbide (VC)-reinforced composites) for wear/corrosion, tailored composition gradients, and strong adhesion. Disadvantages include added surface roughness, dilution control, and thermal distortion. AM adaptations include Taguchi-optimized laser cladding to control carbide fraction and hardness, and hybrid AM enabling functionally graded skins (Dahotre et al., 2016).

7.7.3 Sol-gel and Polymeric Coatings

The principle of sol-gel and polymeric coatings entails the dip/spray/deposition of oxide or polymer precursors, followed by curing to form barrier or functional layers (e.g. bioactive, hydrophobic). Benefits include low-temperature, conformal, and tunable chemistry. Disadvantages include lower mechanical robustness than metallic/ceramic thick coatings; adhesion on rough AM surfaces requires pretreatment. AM adaptations include bioactive sol-gel on medical AM implants, and polymer sealants to improve leak-tightness by porosity sealing (Ramezani et al., 2023).

7.7.4 Hybrid and Multi-step Treatments

Combining mechanical and chemical finishing is often synergistic: roughness reduction by grinding/tumbling, followed by electropolishing to level micro asperities and passivate, concluding with PVD hard coating offers cumulative benefits (fatigue, corrosion, tribology). For internal passages, AFM, followed by chemical passivation, reduces Ra and enhances corrosion resistance. For fatigue-critical parts, sequence grinding → shot peening/laser peening → PVD can deliver smooth, compressively stressed, hard surfaces. For EBM/PBF downskins, laser remelting, followed by light blasting and EP can achieve both pore closure and low Ra (Dahotre et al., 2016; Lee et al., 2020; Ramezani et al., 2023).

In total, the surface engineering toolbox for AM spans subtractive, chemical, thermal, and deposition techniques. The selection of these techniques should be driven by surface function (fatigue vs. tribology vs. corrosion vs. flow), geometry (internal vs. external), alloy, and productivity, often implemented as hybrid sequences to achieve multiple objectives concurrently (Dahotre et al., 2016; Lee et al., 2020; Ramezani et al., 2023).

7.8 IN-PROCESS OR IN-SITU SURFACE MODIFICATION

Integrating surface modification within the AM build addresses surface integrity at its origin, reducing rework and improving consistency. Several strategies have emerged.

7.8.1 In-situ Laser Skin Remelting During PBF and DED

Applying a dedicated contour or skin remelt pass with tuned power, speed or overlap after each layer (or periodically) leverages surface tension-driven leveling to reduce roughness, dissolve near-surface pores, and homogenize chemistry. In laser PBF, a lower energy density per pass avoids deep keyholes, while reflowing asperities; in DED, inter-bead remelting reduces bead ripples and enhances bead overlap. Benefits include reduced Ra on sidewalls and downskins, improved leak-tightness via pore sealing, and a smoother baseline for subsequent finishing. Risks, such as microcracking or distortion, are mitigated via careful parameter maps and preheat management (De Pastre et al., 2022).

7.8.2 In-situ Mechanical Interventions (Sliding Severe Plastic Deformation, Fine Particle Peening)

The layer-wise application of mechanical surface treatments (e.g. sliding severe plastic deformation (SPD), fine particle shot peening) during laser powder-bed builds has been proposed to introduce compressive stress, refine grains, and smooth surfaces contemporaneously. Early studies demonstrate feasibility to tailor internal layer properties and reduce post-processing by deploying surface processing between deposited layers in DMLS AlSi10Mg, suggesting control over internal characteristics and reduced roughness accumulation. Such approaches aim to embed fatigue-beneficial stress states and microstructures as the build progresses (Dahotre et al., 2016).

7.8.3 Dual-material Deposition for Functional Skins

Direct energy deposition and laser cladding enable site-specific overlays during build, e.g. depositing wear-resistant carbide-reinforced composites (VC/M7C3) on load surfaces, or corrosion-resistant skins over structural cores, and achieve surface alloying *in situ*. Taguchi-optimized laser cladding controls carbide fraction, hardness, and dilution, delivering high-hardness skins on steel substrates (martensite + VC + complex carbides) with superior wear resistance. In hybrid systems, a PBF core can be over-clad by DED to create graded surfaces (Dahotre et al., 2016).

7.8.4 Real-time Sensing and Closed-loop Control

The enablers of *in-situ* quality are sensors and feedback. Thermal and optical imaging (e.g. infrared pyrometry and coaxial cameras) monitor melt pool size, temperature, and spatter; acoustic/ultrasonic methods capture defects; structured light or impedance-based metrology can reconstruct layer topography. Systems like CLAMIR (a system of infrared thermography for laser metal deposition) and ThermaViz (a system of multivision pyrometry for melt pool and heat flow) exemplify the commercial real-time control of laser power and feed to stabilize deposition and reduce surface defects. IRISS (a system of electron beam deposition) employs sensor suites and CNC logic to adapt beam power, wire feed, and motion for consistent geometry and microstructure. Inkjet-based systems with machine vision map each layer topography and correct subsequent layers (vision-driven adaptive deposition), reducing stair-stepping and surface error accumulation. These monitoring systems contribute indirectly to improved surface condition by preventing track interruptions, balling, and bead overbuild, limiting roughness sources (Lee et al., 2020).

7.8.5 Contour Scanning Strategies and Hatch Optimization

Surface finish improves with dedicated contour perimeters at lower hatch spacing and speeds, reducing waviness; scan vector optimization (e.g. alternating angles, reduced border hatch) reduces LOF near the edges. Fine layer thickness reduces staircases, but increases the time taken; downskin parameter sets (reduced power and speed, higher hatch overlap) improve the underside finish by curbing particle sintering without fusion lumps (De Pastre et al., 2022).

7.8.6 Plasma/Electrolytic *in-situ* Finishing (Emerging)

Plasma electrolytic polishing and low-temperature plasma treatments could be integrated near-line to smooth and activate surfaces shortly after build, benefitting the adhesion of subsequent coatings or reducing micro asperities before HIP/coating; feasibility depends on the cell design and line integration (Ramezani et al., 2023).

These in-process methods do not obviate the need for finishing in high-end applications, but reduce the depth of material removal required, seal harmful surface porosity before HIP, instill beneficial surface states, and provide more uniform starting surfaces for downstream polishing/coating. Importantly, the closed-loop controls that stabilize melt pools and bead geometry curb the root causes of roughness and surface-connected porosity, making post-processing more predictable and less intrusive (Dahotre et al., 2016; Lee et al., 2020; Ramezani et al., 2023).

7.9 APPLICATIONS

7.9.1 Aerospace Hot-section and Structures

Additively manufactured turbine components (vanes, combustor liners and heat exchangers) demand surfaces that are engineered for fatigue, oxidation, and flow. On nickel superalloys, sequences such as laser remelting (to seal pores and smooth skins) → grit blast/ultrasonic surface modification (to improve coating adhesion) → high-velocity oxygen fuel (HVOF)/plasma-sprayed MCrAlY bond coats and YSZ thermal barrier coatings → finishing grind/peen are used to deliver robust thermal protection and surface integrity. Internal cooling channels are smoothed by AFM or electrochemical processes to reduce pressure drops and enhance cooling; external gas-path surfaces are polished/ground to Ra < 1 μm and peened for fatigue to meet engine efficiency and dwell-fatigue requirements. For airframe and structural brackets in Ti-6Al-4V, shot/laser peening after machining/grinding improves high-cycle fatigue (HCF)/low-cycle fatigue (LCF) endurance and foreign-object damage tolerance, addressing the notch sensitivity of AM Ti due to residual surface artifacts (Dahotre et al., 2016; Lee et al., 2020; Ramezani et al., 2023).

7.9.2 Medical Implants and Instruments

Surface engineering reconciles osseointegration with wear/corrosion control. AM lattice structures provide controlled porosity for bone ingrowth; their trabecular surfaces are cleaned (chemical etch) and optionally coated with bioactive layers (e.g. hydroxyapatite sol-gel) to stimulate osteoblast activity and reduce ion release. Articulating regions (e.g. femoral heads) require ultra-low Ra; mechanical polishing and electropolishing minimize debris generation and improve corrosion resistance through stable passive films. Antimicrobial nanostructured coatings (Ag/Zn-containing) reduce infection risks on exposed surfaces, while plasma nitriding or

DLC coatings improve wear resistance of CoCr/Ti bearing pairs. For dental tools and instruments printed in stainless steels, EP delivers Sa in the tens of nanometers with enhanced tarnish resistance, meeting hygiene and aesthetics standards (Lee et al., 2020; Ramezani et al., 2023).

7.9.3 Energy and Process Equipment

Heat exchangers and reactor components benefit from interior surface smoothing to minimize fouling and pressure losses. AFM reduces Ra in AM Ti and Ni channels, boosting flow efficiency; subsequent chemical passivation stabilizes corrosion behavior. For hydrogen/proton exchange membrane (PEM) devices, fine channel surfaces improve mass transport and reduce parasitic losses; electropolishing and coatings mitigate corrosion in acidic environments. In oil and gas, laser-clad corrosion or erosion-resistant overlays (e.g. Ni-based, carbide-reinforced) on AM flow components resist slurry wear and sour conditions; surface sealing via polymer impregnation enhances leak-tightness in porous AM casings.

7.9.4 Industrial Tooling and Dies

AM conformal-cooled molds or dies reduce cycle times. However, the surface quality of cavity and channel walls dictates molding quality and cooling performance. Internal cooling channels are AFM-finished; cavity surfaces are ground or polished to optical finishes; laser cladding can apply hard-facing alloys to wear regions. Ultrasonic nanocrystalline surface modification before plasma spray improves ceramic coating adhesion on tool steels, enhancing wear resistance and coating life; post-spray peening and polishing tune surface topography for polymer release (Dahotre et al., 2016).

7.9.5 Automotive Power Train and Fuel Systems

For AM manifolds, housings, and gear components, surface finishing via tumbling + peening + EP sequences provide Ra reductions and surface compressive stresses to combat micro-pitting and scuffing. PVD coatings (CrN/DLC) on machined tooth flanks reduce friction and wear; sealing faces are lapped to prevent leaks. AM injector nozzles exploit EP to smooth micro-channels and outlets, improving spray atomization and reducing coking, thereby sustaining emissions and efficiency (Lee et al., 2020; Ramezani et al., 2023).

Collectively, these cases underline the role of surface engineering in translating geometrical freedom into functional reliability: finishing and coatings elevate fatigue, wear, corrosion, flow, and sealing performance, ensure regulatory conformance, and expand AM adoption in safety-critical sectors.

7.10 CHARACTERIZATION AND METROLOGY

Effective surface engineering depends on rigorous characterization, linking surface attributes to performance.

7.10.1 Surface Profilometry

Contact stylus and optical profilometry quantify roughness parameters (Ra, Rz, Sa, Sz), waviness, and areal texture. For AM, areal (3D) metrics better capture clustered particle nodules and terrace morphologies; the scan orientation should capture the principal lay. The profilometry baseline informs finishing targets (e.g. PBF Ra ~9–50 μm down to <1 μm) and process efficacy (Lee et al., 2020).

7.10.2 Scanning Electron Microscopy

Scanning electron microscopy (SEM) provides morphological insights into partially fused particles, recast layers, balling droplets, and surface-connected porosity. Fractography identifies crack initiation at LOF or surface pores; cross-sections reveal coating adhesion, diffusion layers (e.g. nitriding), and remelted layers post-laser smoothing (Brennan et al., 2021; Ramezani et al., 2023).

7.10.3 Atomic Force Microscopy and Scanning Tunneling Microscopy

Atomic force microscopy (AFM) quantifies nanometer-scale roughness and reveals polishing artifacts or nanoscale textures, influencing protein/cell interactions or tribological film formation. Scanning tunneling microscopy (STM) assists in conductive surfaces for atomic-scale texture when relevant (Ramezani et al., 2023).

7.10.4 X-ray Diffraction

X-ray diffraction (XRD) identifies phase composition, residual stress ($\sin 2\psi$) states, and coating phases (e.g. TiN, CrN; amorphous carbon). In AM, XRD evaluates microstructural changes after laser remelting or heat treatments, and confirms diffusion-hardening surface layers (expanded austenite).

7.10.5 Nanoindentation and Microhardness

Indentation profiles map hardness and modulus gradients introduced by peening, nitriding, or coatings; nanoindentation isolates thin film properties. For composite laser clads (e.g. VC in martensitic matrices), micro hardness correlates with carbide fraction and wear resistance (Dahotre et al., 2016; Ramezani et al., 2023).

7.10.6 Tribology Testing

Pin-on-disk, reciprocating wear, and scuffing tests evaluate friction coefficients, wear rates, and transfer films under relevant lubrication; tests benchmark finishing or coating benefits. AM surfaces show reduced wear and friction after smoothing or coating; surface texturing can be tuned for lubricant retention (Ramezani et al., 2023).

7.10.7 Corrosion Testing

Potentiodynamic polarization, electrochemical impedance spectroscopy (EIS), and salt spray tests quantify corrosion rates and pitting potentials; EP surfaces typically show lower current densities and more noble behavior due to improved passive films. Crevice tests are relevant for surface-connected porosity; coatings are assessed for barrier performance (Ramezani et al., 2023).

7.10.8 Non-destructive Evaluation

X-ray computed tomography detects near-surface and internal defects; micro-computed tomography (CT) visualizes pore intersections with surfaces and allows pore sealing efficacy checks after remelting or impregnation. Thermal/infrared imaging and ultrasound can detect surface-breaking defects in coated parts (Brennan et al., 2021; De Pastre et al., 2022).

7.10.9 Process Monitoring Data

Melt pool thermal signatures, optical topography maps (layer-wise), and impedance-based sensing feed into surface outcome prediction and corrective actions (e.g. adaptive hatching); this data augments traditional metrology for in-process surface control (Lee et al., 2020).

Correlating measures to performance is key: for fatigue, link the surface roughness parameter (Rz) and defect distributions (size, surface connectivity) to the stress-number (S-N) behavior; for corrosion, link EP-induced roughness/pit density to polarization; for tribology, tie Ra/Sa and microstructure to friction/wear. This measurement-to-function mapping closes the loop between finishing parameters, achieved surface states, and service outcomes.

7.11 FUTURE TRENDS AND RESEARCH DIRECTIONS

7.11.1 AI and Machine Learning for Surface Optimization

Data-driven models that have been trained on process-structure-surface datasets (process parameters, melt pool signals, layer topography, roughness, defect maps, and performance metrics) will recommend *in-situ* contour or remelt parameters, scan strategies, and post-processing sequences to hit target Ra/defect thresholds with minimal finishing burden. Active-learning frameworks that use prioritized sampling have been proposed to improve AM model selection under data scarcity, suggesting efficient paths to predictive surface-quality models. Integrated with in-process sensing (infrared, vision), machine learning can drive closed-loop adjustments (power, speed) to suppress balling, reduce spatter deposition, and stabilize bead geometry, thus preventing surface defects at the source (Lee et al., 2020, Ramos et al., 2025).

7.11.2 Smart/Functional Coatings

Coatings that indicate wear (color/fluorescence changes), inhibit microbes (bioactive/antimicrobial layers), or self-clean (hydrophobic/photocatalytic) are likely to proliferate on AM parts with complex geometries. Nanostructured and multilayer PVD/CVD stacks deliver tailored hardness or strain accommodation; sol-gel-derived bioactive films and antimicrobial nanocomposites target medical AM devices; self-healing barrier coatings and lubricious films extend the service of AM tribology pairs. Laser cladding-based graded overlays and *in-situ* synthesized phases during SEAM further expand site-specific functionality (Dahotre et al., 2016; Ramezani et al., 2023).

7.11.3 Sustainability, Scalability, and Economics

Surface finishing can dominate cost and cycle time in AM. Research must address media/chemical recycling for mass finishing and EP; low-energy, low-waste alternatives (e.g. plasma electrolytic polishing), and process chains that minimize material removal via in-process smoothing. Life cycle assessments should capture energy and emissions for finishing steps: finishing choices influence AM's sustainability claims. Standards development (ISO/ASTM) around surface finishing practices for AM will accelerate industrial adoption with consistent quality benchmarks (e.g. ISO/ASTM 52904 practice for PBF critical applications) and test artifacts for geometric/surface capability (Lee et al., 2020).

7.11.4 Industry 4.0 Integration, Inline Monitoring, Digital Twins

Digital twins of surface formation, coupled with on-machine sensing, will forecast as-built roughness and defect likelihood per feature, enabling proactive parameter adjustments and the automated selection of minimal finishing recipes. Systems such as CLAMIR, ThermaViz, and IRISS illustrate the blueprint for sensor-driven control; expanding to fine-grained, per-feature surface predictions closes the loop from design intent (surface function) to manufacturing execution. Vision-guided layer correction (Inkbit) and impedance-based layer quality tracking further herald the in-line assurance of surface outcomes. Linking metrology data back to design databases enhances design-for-surface rules across AM modalities (De Pastre et al., 2022).

7.11.5 Cross-cutting Materials Research

The surface engineering of AM-specific microstructures (cellular, textured) requires tailored treatments to avoid embrittlement (e.g. nitriding without alpha-case in Ti), mitigates sensitization in stainless steels, and controls phase stability in nickel alloys. Low-temperature plasma treatments, EP chemistries for AM surfaces, and coatings that are tolerant of AM substrate roughness and residual stress states are priorities. The *In-situ* synthesis (SEAM) of functionally graded skins during DED/cladding expands possibilities for "designing the surface as you build" (Dahotre et al., 2016; Ramezani et al., 2023).

7.12 CONCLUSION

Surface engineering is the essential complement to AM by converting as-built, particle-decorated, layer-stepped artifacts into application-grade components by removing harmful features, instilling beneficial stress states and chemistries, sealing pores, and adding functional skins. Across sectors—biomedical, aerospace, automotive, and energy—surface condition governs fatigue life, wear/friction behavior, corrosion resistance, fluid transport, and sealing. The techniques presented show that, while AM processes enable unprecedented geometries, intrinsic surface challenges—roughness, surface-connected porosity, residual stresses, and near-surface microstructural anomalies—must be systematically addressed by mechanical, chemical, thermal/plasma, and coating technologies, often sequenced as hybrid chains to achieve multiple objectives efficiently.

Some core challenges remain in the whole process of surface engineering, like finishing internal channels and lattice interiors at scale, minimizing finishing cost and reducing the environmental footprint, achieving deterministic surface states on complex features, and integrating in-process smoothing with robust and standardized quality control. Promising innovations include *in-situ* laser remelting, layer-wise mechanical treatments, sensor-driven closed-loop deposition control, and site-specific overlays via SEAM. On the finishing side, advanced AFM and electrochemical methods for internal surfaces, plasma electrolytic and EP chemistries that are tuned to AM microstructures, and nanostructured multifunctional coatings are expanding the surface engineer's toolkit.

Looking forward, AI-assisted process planning, tied to in-line sensing and digital twins, will enable "design-to-surface" workflows that deliver target roughness and surface integrity with minimal rework. Standardized artifacts and practices will solidify confidence for critical applications. Ultimately, surface engineering is the lever that unlocks AM's full potential—ensuring that geometry freedom is matched by surface performance—so that AM can meet the most demanding service environments with reliability and efficiency.

REFERENCES

Abdulhameed, O., Al-Ahmari, A., Ameen, W., & Mian, S. H. (2019). Additive manufacturing: Challenges, trends, and applications. *Advances in Mechanical Engineering*, 11(2), 1–27. https://doi.org/10.1177/1687814018822880

Bikas, H., Stavropoulos, P., & Chryssolouris, G. (2015). Additive manufacturing methods and modelling approaches: A critical review. *International Journal of Advanced Manufacturing Technology*, 83, 389–405. https://link.springer.com/article/10.1007/s00170-015-7576-2.

Brennan, M. C., Keist, J. S., & Palmer, T. A. (2021). Defects in metal additive manufacturing processes. *Journal of Materials Engineering and Performance*, 30(7), 4808–4818. https://doi.org/10.1007/s11665-021-05919-6

Chen, Y., Peng, X., Kong, L., Dong, G., Remani, A., & Leach, R. (2021). Defect inspection technologies for additive manufacturing. *International Journal of Extreme Manufacturing*, 3(2), 022002. https://doi.org/10.1088/2631-7990/abe0d0

Dahotre, N. B., Vora, H. D., & Boesl, B. (2016). Additive manufacturing via surface engineering. *JOM*, 68(7), 1759–1761. https://doi.org/10.1007/s11837-016-1955-5

de Pastre, M.-A., Quinsat, Y., & Lartigue, C. (2022). Effects of additive manufacturing processes on part defects and properties: A classification review. *International Journal on Interactive Design and Manufacturing (IJIDeM)*, 16, 1471–1496. https://doi.org/10.1007/s12008-022-00839-8

Lee, J.-Y., Nagalingam, A. P., & Yeo, S. H. (2020). A review on the state-of-the-art of surface finishing processes and related ISO/ASTM standards for metal additive manufactured components. *Virtual and Physical Prototyping*, 16, 68–96. https://doi.org/10.1080/17452759.2020.1830346

Prakash, K. S., Nancharaih, T., & Subba Rao, V. V. (2018). Additive manufacturing techniques in manufacturing—An overview. *Materials Today: Proceedings*, 5, 3873–3886. https://doi.org/10.1016/j.matpr.2018.06.219

Rajaguru, K., Karthikeyan, T., & Vijayan, V. (2019). Additive manufacturing—State of art. *Materials Today: Proceedings*, 21, 628–633. https://doi.org/10.1016/j.matpr.2019.06.728

Ramezani, M., Mohd Ripin, Z., Pasang, T., & Jiang, C.-P. (2023). Surface engineering of metals: Techniques, characterizations and applications. *Metals*, 13(7), 1299. https://doi.org/10.3390/met13071299

Ramos, A., Angel, V. G., Siqueiros, M., Sahagun, T., Gonzalez, L., & Ballesteros, R. (2025). Reviewing additive manufacturing techniques: Material trends and weight optimization possibilities through innovative printing patterns. *Materials*, 18(6), 1377. https://doi.org/10.3390/ma18061377

Rasiya, G., Shukla, A., & Saran, K. (2021). Additive manufacturing—A review. *Materials Today: Proceedings*, 47, 6896–6901. https://doi.org/10.1016/j.matpr.2021.05.181

Rivera, F. , & Arciniegas, A. J. (2020). Additive manufacturing methods: Techniques, materials, and closed-loop control applications. *International Journal of Advanced Manufacturing Technology*, 109, 17–31. https://link.springer.com/article/10.1007/s00170-020-05663-6

Zhang, Y., Jarosinski, W., Jung, Y.-G., & Zhang, J. (2018). Additive manufacturing processes and equipment. In Zhang, J. & Jung, Y-G (Eds.), *Additive Manufacturing* (pp. 39–51). Elsevier. https://doi.org/10.1016/B978-0-12-812155-9.00002-5

8 Sustainable Development with Additive Manufactured Materials

Prashant Kumar Singh, Sanjay Yadav, Siddhartha Sharma, Naveen Mani Tripathi, Navdeep Sharma Dugala, and Rahul Nadda

8.1 INTRODUCTION

Additive manufacturing (AM), more commonly referred to as 3D printing, has been an evolving technology that can change the manufacturing model of sustainable industries. Through the layer-by-layer addition of components from computer-aided design (CAD) information, AM minimizes material loss—more than 90% in some cases relative to subtractive techniques—and enables the production of complex, weightless geometries that are capable of saving energy and reducing emissions, particularly in energy-conscious industries such as the aerospace and automotive industries (Shah et al., 2024). Yet, not all AM processes are created equal: energy used in selective laser melting (SLM) and direct energy deposition (DED) processing can be orders of magnitude more, and stand-by time or follow-on processes such as electrical discharge machining (EDM) in metal AM will have a tendency to control the life cycle energy and carbon footprint (Rahmani et al., 2025). Concurrently with geometry-based efficiency, green feedstock innovation has now started to become a key focus area in greenifying AM. Recently available breakthrough research has shown direct 3D printing of bio-composites from mycelium based on waste streams (e.g. cardboard or spent coffee grounds) with a fungal inoculum addition, enabling fully compostable components to be fabricated without molds or tooling (Luo et al., 2025). Related work proposes methods of developing structurally resilient mycelium architectures from agro-industrial by-products, achieving hydrophobic, lightweight, and biodegradable material, which is useful for packaging, insulation, and even architecture.

Concurrent research has explored novel eco-polymers and recycled plastics such as post-consumer polylactic acid (PLA) and acrylonitrile butadiene styrene (ABS) for 3D printing, exhibiting uneven mechanical performance, but hinting at the possibility of circular supply chain development through standardization and life cycle assessment (LCA) systems (Pinho et al., 2020). Increasingly, literature also emphasizes the prospect of AM in decentralized, local production systems with lower transport-related

DOI: 10.1201/9781003643531-8

emissions and enhanced supply-chain resilience. Overviews published in 2024/25 highlight AM as key to translating Industry 5.0/6.0 visions into reality—smart, human-oriented, energy-efficient production on the basis of the Internet of Things (IoT) and artificial intelligence (AI)-enhanced process optimization—where on-demand manufacturing enables repair, remanufacturing, and circularity (Rojek et al., 2025). Construction-specific systematic reviews of AM also affirm that, despite the promise, effective greenhouse gas (GHG) mitigation hinges on the process selection, feedstock source, scale, and standardization of sustainability metrics (Oladunni et al., 2025).

Despite the dramatic advances, however, there are some important gaps. Energy-intensive operation remains an issue in metal-based AM, and even more so in the case of parts made using laser-based systems where idle time, post-processing, and machine-embodied energy play a significant role in life cycle impact (Cozzolino et al., 2025). In addition, while biocomposite materials like mycelium composites and recycled polymers have profound environmental benefits, their large-scale application involves strong, normalized testing programs, encompassing entire product life cycles, mechanical toughness, and contamination mitigation through growth cycles (Luo et al., 2025). Research agendas point out the need to create sustainable feedstocks, maximize machine energy use efficiency, and normalize cooperation in LCA and circular economy metrics to guarantee that AM advancements bring the anticipated environmental impacts (Kellens et al., 2017). In combination, geometry-driven material reduction in AM, new bio-based and recycled feedstocks, and the transition to distributed, digitally connected manufacturing environments place AM at the forefront of sustainable production. Its potential is achieved by optimizing energy efficiency, scaling up eco-feedstock innovation, and creating standardized life cycle and sustainability measurements—precisely the challenges confronting this work (Ribeiro et al., 2020).

The sustainability aspects should be considered in the early stages of the product development process, because once the product reaches the manufacturing stage, the opportunities to modify the product become minimal (Singh, 2022). Therefore, the focus in this study is to develop a metric that designers can use in the earlier design stages to evaluate a product's sustainability potential.

8.2 METHODOLOGY

The methodology adopted in this study has been presented through a flowchart as shown in Figure 8.1.

Different sustainability criteria included in the study are discussed below.

8.2.1 Cost of Material

The cost of material constitutes a significant portion of the overall expense in AM, particularly when considering sustainability implications. Material costs, along with the initial investment in hardware and software, are major drivers in AM (Yang & Li, 2018). Despite decreasing costs of AM systems, material expenses remain a crucial

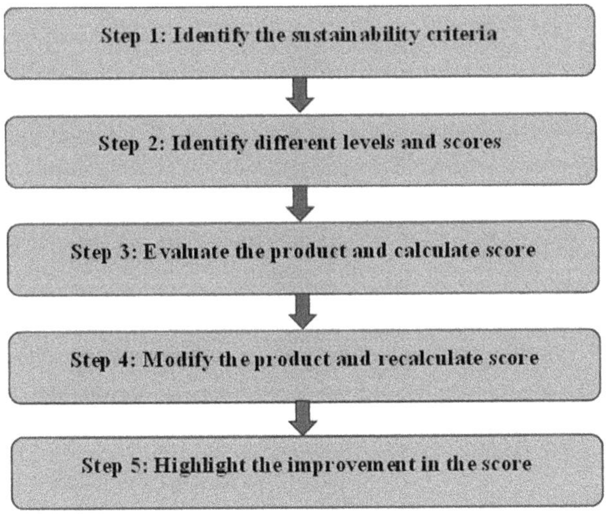

FIGURE 8.1 Flowchart of the proposed methodology.

factor influencing adoption of the technology across various industries (Thomas, 2016). The selection of materials profoundly impacts not only the economic viability, but also the environmental footprint of the entire manufacturing process (Javaid et al., 2021). Efforts to reduce costs include optimizing material usage, exploring alternative materials, and implementing efficient recycling strategies. The environmental effects related to material production, processing, and disposal cannot be ignored (Kellens et al., 2017). Therefore, a comprehensive approach is needed that considers both economic and ecological elements to promote long-term sustainability in additive manufacturing (Ford & Despeisse, 2016).

8.2.2 Recyclability of Material

The convergence of AM and sustainable practices has brought material recyclability to the forefront, presenting both opportunities and challenges. The pursuit of sustainability in AM is gaining traction as industries strive to minimize their ecological impact (Javaid et al., 2021). Additive manufacturing offers a transformative approach that can revolutionize environmental implications throughout the entire product life cycle, enhancing the potential for sustainability (Nadagouda et al., 2020).

8.2.3 Source of Energy

Additive manufacturing can significantly enhance sustainability in energy systems by enabling lightweight and topology-optimized components, powered by renewable sources. When AM facilities run on solar or wind energy, carbon footprints decline—e.g. carbon emissions for aluminum parts via laser powder bed fusion (PBF) drop by nearly 50%.

8.2.4 Post-processing

Post-processing is pivotal for AM's life cycle sustainability. Advanced systems recover unused powder and condition support media—minimizing material waste and improving circularity in selective laser sintering (SLS) and multi-jet fusion (MJF) workflows. Automated, software-driven systems reduce energy and chemical usage, optimize detergent usage, and streamline finishing while cutting labor.

8.2.5 Time Required to Build the Product

Recent studies show that build time in AM—encompassing design, printing, and post-processing—directly influences both energy use and carbon emissions, making it a primary focus in sustainable AM research. Industry reports indicate that typical AM production lead times range from a few hours (small stereolithography (SLA)/ fused deposition modeling (FDM) prototypes) to multiple days for medium-sized or metal parts—e.g. one to three days for plastic parts; five to ten days for metal end-use components; and seven to 14 days for multi-material assemblies, depending on post-processing and complexity. Overall, reducing build time through process optimization remains fundamental to advancing the sustainability of AM operations, offering both direct reductions in energy consumption and indirect benefits via material efficiency, lightweight design, and localized production models (Nyamekye et al., 2024).

8.2.6 Transport Mode to Supply Materials

Recent research indicates that transport modes for supplying AM materials significantly influence sustainability outcomes. AM relies on compact raw material inputs (powders or filaments), enabling consolidated bulk transport, often by rail or sea, instead of the frequent shipment of finished parts, which reduces emissions per mass shipped (Naghshineh & Carvalho, 2025).

8.2.7 Number of Components in the Assembly

Reducing the component count through part consolidation in AM significantly enhances sustainability by lowering material use, assembly energy, and supply chain complexity. Design for Additive Manufacturing (DfAM) principles enable radical consolidation (GE Aviation, for instance, reduced a helicopter engine from 900 to 16 parts), cutting embodied energy and logistics emissions. Fewer components mean fewer support structures, less post-processing, and improved recyclability, reducing waste and human toxicity risks. Topology optimization and self-supporting geometries further minimize part segmentation and build time, amplifying life cycle CO_2 and energy gains (Jung et al., 2023).

8.2.8 Ease of Use/Ergonomics

Optimizing ergonomic design in AM workspaces improves sustainability by reducing operator fatigue and errors, thereby enhancing throughput and lowering waste from misprints. Integrating human-centered design and digital human modelling in AM post-processing and printing workflows promotes more efficient operations and safer operator interactions, and reduces energy-intensive reworks. Ergonomic usability supports life cycle sustainability by minimizing failed builds and encouraging the adoption of eco-friendly materials like PLA, which emit fewer volatile organic compounds (VOCs) and particulates when printed under controlled, user-friendly conditions. This leads to more resilient, sustainable AM systems (Hasanain, 2024).

8.2.9 Maintainability/Reparability

Additive manufacturing's strength in repairability and maintainability enhances sustainability by extending product lifetimes and reducing waste. Emerging AM repair workflows—such as Pratt & Whitney's laser powder AM repair for turbine engines—can cut turnaround time by over 60%, enable component reuse, and lessen dependency on new material supply chains. Research frameworks integrate on-demand, locally printed replacement parts to support rapid repair systems, significantly reducing logistics emissions and material waste. Using infinitely recyclable resins and distributed filament recycling (e.g. recyclebots) further increases life cycle circularity and resource efficiency (Shah et al., 2024).

8.2.10 Durability

Additive manufacturing's ability to produce durable, high-performance parts enables a longer lifespan and more sustainable outcomes. Materials like high entropy alloys (HEAs), manufactured by laser-based AM, exhibit improved strength and ductility, enhancing resistance to wear and fracture. SLM produces dense, precise metal components that, despite higher energy input, yield lightweight parts—such as aerospace blades—that reduce operational fuel use and offset production emissions across life cycles. A sustainability review highlights that multi-material, optimized lattice architectures extend structural durability, while minimizing material waste (Shah et al., 2024).

The different levels of each sustainability criteria and the corresponding sustainability scores are presented in Table 8.1. These levels and scores are determined in consultation with a team of five members. It includes all the authors and an industry expert working in the area of AM. A higher score indicates a more sustainable product.

The applicability of the proposed approach has been demonstrated with the help of a case study. This case study includes the design and manufacturing of a spur gear through an AM approach. The functionalities of the gear that have been considered are its application in washing machines and automated teller machines (ATMs), i.e. low- to medium-load applications. The case study is carried out under two scenarios. These scenarios are discussed in the results section.

TABLE 8.1
Sustainability Criteria with Levels and Corresponding Scores

Sr. No.	Criteria	Level	Score
1	Cost of material	Low (such as polymers)	**10**
		Medium (metals and alloys)	**5**
		High (composites and biomaterials)	**0**
2	Recyclability of material	Low (0–10%)	**0**
		Medium (11–50%)	**5**
		High (more than 50%)	**10**
3	Source of energy	Renewable (solar, wind, etc.)	**10**
		Mixed	**5**
		Fossil fuels (coal, diesel, etc.)	**0**
4	Post-processing	Low (support removal or cleaning)	**10**
		Medium (painting)	**5**
		High (machining)	**0**
5	Time required to build the product	Low (few minutes)	**10**
		Medium (few hours)	**5**
		High (many hours or days)	**0**
6	Transport mode to supply materials	Air	**0**
		Road/rail	**5**
		Ship	**10**
7	Number of components in the assembly	Standalone	**10**
		Few components (2 to 3)	**5**
		Multiple components	**0**
8	Ease of use/ergonomics	Low	**0**
		Medium	**5**
		High	**10**
9	Maintainability/reparability	Low	**0**
		Medium	**5**
		High	**10**
10	Durability	Low	**0**
		Medium	**5**
		High	**10**

8.3 RESULTS AND DISCUSSION

The results are presented under two different scenarios as discussed below.

8.3.1 Scenario 1

In this case, the gear is manufactured utilizing the SLM approach of AM. The metallic powder used is Stainless Steel 316L. The application of the proposed metrics on the manufactured metal gear is presented in Table 8.2.

TABLE 8.2
Results of Scenario 1

Metallic spur gear
Material: Stainless Steel 316L
Method: SLM

Sr. No.	Criteria	Level	Score
1	Cost of material	Medium (metals and alloys)	5
2	Recyclability of material	Medium (11–50%)	5
3	Source of energy	Fossil fuels (coal, diesel etc.)	0
4	Post-processing	High (machining)	0
5	Time required to build the product	Medium (few hours)	5
6	Transport mode to supply materials	Road/rail	5
7	Number of components in the assembly	Standalone	10
8	Ease of use/ergonomics	Medium	5
9	Maintainability/reparability	High	10
10	Durability	High	10
Total score			**55**

8.3.2 Scenario 2

In this case, the gear is manufactured utilizing the FDM approach of AM. The plastic material used is ABS. The application of the proposed metrics on the manufactured plastic gear is presented in Table 8.3.

A comparison of Scenario 1 with Scenario 2 indicates that the selection of a plastic gear for low- to medium-load applications is 31% more sustainable than that of a metallic gear.

TABLE 8.3
Results of Scenario 2

Plastic spur gear
Material: ABS
Method: FDM

Sr. No.	Criteria	Level	Score
1	Cost of material	Low (such as polymers)	10
2	Recyclability of material	High (more than 50%)	10
3	Source of energy	Renewable (solar, wind, etc.)	10
4	Post-processing	Low (support removal or cleaning)	10
5	Time required to build the product	Low (few minutes)	10
6	Transport mode to supply materials	Road/rail	5
7	Number of components in the assembly	Standalone	10
8	Ease of use/ergonomics	High	10
9	Maintainability/reparability	Low	0
10	Durability	Medium	5
Total score			**80**

8.4 CONCLUSION

The current study focuses on the development of a metric to assess the sustainability potential of a product before its manufacture. The developed metric consists of a number of sustainability criteria, which are further classified according to different levels. Each level has been assigned a sustainability score. A higher score indicates better sustainability. The suggested metric helps develop different scenarios of a product by altering the different levels of sustainability criteria. Thus, the resulting sustainability scores corresponding to different scenarios help the designers and manufacturers select the best scenario (i.e. one having the highest score).

The developed metric is still in its nascent stage and can be further enhanced by adding more relevant sustainability criteria. The criteria are also qualitative in nature, and are hence only indicative. They do not measure the actual environmental impacts of the different scenarios.

REFERENCES

Cozzolino, E., Papa, I., & Astarita, A. (2025). Energy assessment of powder bed fusion additive manufacturing processes at industrial scale: Experiments and simulations. *Journal of Materials Engineering and Performance*, *34*(15), 15315–15323. https://doi.org/10.1007/s11665-025-11007-w

Ford, S., & Despeisse, M. (2016). Additive manufacturing and sustainability: An exploratory study of the advantages and challenges. *Journal of Cleaner Production*, *137*, 1573–1587. https://doi.org/10.1016/j.jclepro.2016.04.150

Hasanain, B. (2024). The role of ergonomic and human factors in sustainable manufacturing: A review. *Machines*, *12*(3), 159. https://doi.org/10.3390/machines12030159

Javaid, M., Haleem, A., Singh, R. P., Suman, R., & Rab, S. (2021). Role of additive manufacturing applications towards environmental sustainability. *Advanced Industrial and Engineering Polymer Research*, *4*(4), 312–322. https://doi.org/10.1016/j.aiepr.2021.07.005

Jung, S., Kara, L. B., Nie, Z., Simpson, T. W., & Whitefoot, K. S. (2023). Is additive manufacturing an environmentally and economically preferred alternative for mass production? *Environmental Science & Technology*, *57*(16), 6373–6386. https://doi.org/10.1021/acs.est.2c04927

Kellens, K., Mertens, R., Paraskevas, D., Dewulf, W., & Duflou, J. R. (2017). Environmental impact of additive manufacturing processes: Does AM contribute to a more sustainable way of part manufacturing? *Procedia CIRP*, *61*, 582–587. https://doi.org/10.1016/j.procir.2016.11.153

Luo, D., Yang, J., & Peek, N. (2025). 3D-printed mycelium biocomposites: Method for 3D printing and growing fungi-based composites. *3D Printing and Additive Manufacturing*, *12*(2), 98–111. https://doi.org/10.1089/3dp.2023.0342

Nadagouda, M. N., Ginn, M., & Rastogi, V. (2020). A review of 3D printing techniques for environmental applications. *Current Opinion in Chemical Engineering*, *28*, 173–178. https://doi.org/10.1016/j.coche.2020.08.002

Naghshineh, B., & Carvalho, H. (2025). Towards a practice-based framework for supply chain resilience in the context of additive manufacturing technology adoption. *International Journal of Computer Integrated Manufacturing*, *38*(3), 335–361. https://doi.org/10.1080/0951192X.2024.2333011

Nyamekye, P., Lakshmanan, R., Tepponen, V., & Westman, S. (2024). Sustainability aspects of additive manufacturing: Leveraging resource efficiency via product design optimization and laser powder bed fusion. *Heliyon*, *10*(1), e23152. https://doi.org/10.1016/j.heliyon.2023.e23152

Oladunni, O. J., Lee, C. K. M., Ibrahim, I. D., & Olanrewaju, O. A. (2025). Advances in sustainable additive manufacturing: A systematic review for construction industry to mitigate greenhouse gas emissions. *Frontiers in Built Environment*, *11*, 1535626. https://doi.org/10.3389/fbuil.2025.1535626

Pinho, A. C., Amaro, A. M., & Piedade, A. P. (2020). 3D printing goes greener: Study of the properties of post-consumer recycled polymers for the manufacturing of engineering components. *Waste Management*, *118*, 426–434. https://doi.org/10.1016/j.wasman.2020.09.003

Rahmani, R., Bashiri, B., Lopes, S. I., Hussain, A., Maurya, H. S., & Vilu, R. (2025). Sustainable additive manufacturing: An overview on life cycle impacts and cost efficiency of laser powder bed fusion. *Journal of Manufacturing and Materials Processing*, *9*(1), 18. https://doi.org/10.3390/jmmp9010018

Ribeiro, I., Matos, F., Jacinto, C., Salman, H., Cardeal, G., Carvalho, H., Godina, R., & Peças, P. (2020). Framework for life cycle sustainability assessment of additive manufacturing. *Sustainability*, *12*(3), 929. https://doi.org/10.3390/su12030929

Rojek, I., Mikołajewski, D., Kopowski, J., Bednarek, T., & Tyburek, K. (2025). Energy efficiency and sustainability of additive manufacturing as a mass-personalized production mode in industry 5.0/6.0. *Energies*, *18*(13), 3413. https://doi.org/10.3390/en18133413

Shah, H. H., Tregambi, C., Bareschino, P., & Pepe, F. (2024). Environmental and economic sustainability of additive manufacturing: A systematic literature review. *Sustainable Production and Consumption*, *51*, 628–643. https://doi.org/10.1016/j.spc.2024.10.012

Singh, P. K. (2022). *Understanding and Developing Ecodesign Support for Environmentally Conscious Product Development* [PhD Thesis]. Indian Institute of Technology Ropar.

Thomas, D. (2016). Costs, benefits, and adoption of additive manufacturing: A supply chain perspective. *The International Journal of Advanced Manufacturing Technology*, *85*(5–8), 1857–1876. https://doi.org/10.1007/s00170-015-7973-6

Yang, Y., & Li, L. (2018). Cost modeling and analysis for mask image projection stereolithography additive manufacturing: Simultaneous production with mixed geometries. *International Journal of Production Economics*, *206*, 146–158. https://doi.org/10.1016/j.ijpe.2018.09.023

9 Additive Manufacturing in Developing Markets

Balwant Singh, Jaspreet Singh, Bikramjeet Singh, and Madhusudan Baghel

9.1 INTRODUCTION

Additive manufacturing (AM) is a game changer for the Third World, where traditional production is sometimes impracticable or impossible due to a lack of infrastructure, financial constraints, and technological restrictions. The first 3D-printed post office in India, for instance, opened its doors in Bengaluru's Cambridge Layout on August 18, 2023. The creative endeavor was more than just a technological showcase; it was completed in 43 days and cost between 30 and 40% less than traditional construction methods. This practical study demonstrates AM's revolutionary potential for infrastructure development, with implications for exploring more extensive applications in developing countries.

Because AM can create complex parts without the significant capital outlay often required for factories, custom molds, and drawn-out supply networks, it is particularly revolutionary for developing countries. Smaller businesses can create high-value goods locally thanks to AM's decentralization, which lowers transportation costs and the need for imports in areas where such infrastructure might be prohibitively expensive. According to Table 9.1, the post office's exceptionally short construction timeframe served as a striking example of AM's ability to shorten product development cycles (Research and Markets, 2022).

With a current valuation of USD 450 million and a remarkable compound annual growth rate of 30%, India's AM market is proof of that disruptive potential. In line with worldwide trends, developing markets are increasingly using AM to encourage its use by facilitating its specialized incubation institutes. This is not exclusive to India; according to a 2023 World Bank assessment, the country's healthcare and education sectors will be the main drivers of the strong AM market in sub-Saharan Africa's strong 21% compound annual growth rate (CAGR) by 2030 (Hou, 2025; Tegethoff et al., 2025). This will diversify their economies, and actively promote the government's "Make in India" campaign.

DOI: 10.1201/9781003643531-9

TABLE 9.1
Overview of Additive Manufacturing in Developing Markets

Aspect	Description
Definition	Layer-by-layer production from digital models
Merits	Reduced waste, customization, decentralized manufacturing
Relevance	Overcomes infrastructural and financial barriers
Case study	India's 3D-printed post office, 43 days, 30–40% cost savings

TABLE 9.2
Economic Challenges and AM Solutions

Challenge	AM solution	Example
Infrastructure	Decentralized production reduces logistics needs	Bengaluru post office built in 43 days
Capital costs	Low-cost FDM printers enable SME adoption	India's USD 450 million AM market, 30% CAGR
Import dependency	Local filament production from recycled materials	Reflow's PET filament in Pune
Skills gap	Training programs for AM operation	NCAM training 1000 technicians annually

9.2 ECONOMIC AND INDUSTRIAL LANDSCAPE OF DEVELOPING MARKETS

The economies of the majority of developing nations, including India, are struggling. Their economies are still largely reliant on low-value industries and agriculture, with manufacturing, which shockingly makes up only 16% of India's gross domestic product (GDP) and barely 10 to 15% in sub-Saharan Africa, making a negligible contribution. Deteriorated infrastructure, high logistics costs, erratic energy supplies, and a significant reliance on imports that drain limited foreign investment are just a few of the problems these nations face (Handfield et al., 2022; Rashid et al., 2020). AM's tale becomes more interesting at this point (Table 9.2) (Haleem et al., 2020).

9.2.1 Global Success Stories

These problems are not limited to India. By using fab labs to produce parts to fit their farm equipment on demand, small business owners in Nairobi, Kenya, have dramatically reduced downtime and costs for farmers (Baumers et al., 2017; Haleem et al., 2020). The 2013 study found that AM has the revolutionary potential to use rapid prototyping to accelerate everything from product development to market introduction (Lanz et al., 2018). Consider Indian startups like 3D Incredible, which uses AM to produce drone parts. To illustrate how technology can be used to diversify the economy, the startup has reduced lead times from months to just weeks (Lydon et al., 2019).

9.2.2 Creating Jobs and Building Futures

Since 2018, AM has helped South Africa's Centre for Rapid Prototyping and Manufacturing (CRPM) generate 500 positions in the healthcare and aerospace industries.

These families have steady incomes, and communities are developing technological capabilities; they are more than just statistics. They demonstrate how AM may promote growth that is truly inclusive across developing nations, creating opportunities that traditional manufacturing techniques might not have been able to provide (Mellor et al., 2014).

9.3 ADDITIVE MANUFACTURING TECHNOLOGIES AND RELEVANCE

According to Table 9.3, the AM technologies are binder jetting (BJ), selective laser sintering (SLS), stereolithography (SLA), and fused deposition modeling (FDM).

- FDM: Used with thermoplastics like polylactic acid (PLA) and acrylonitrile butadiene styrene (ABS), this low-cost method is ideal for construction and prototyping. To construct a bell-shaped building without pillars, the Bengaluru post office used a COBOD BOD2 printer and a fast-hardening concrete mixture that had been tested and approved by IIT Madras.
- SLA: This high-resolution, photopolymer-based material is appropriate for dentistry and medical uses. SLA is used by Anatomiz3D in India to create surgical models.
- SLS: This process is expensive, and uses metals or nylon for industrial and aerospace components. It is less prevalent in underdeveloped nations.
- Binder jetting: This is expensive for castings made of ceramic and metal materials.

TABLE 9.3
Additive Manufacturing Technologies in Developing Markets

Technology	Cost	Resolution	Speed	Materials	Application
FDM	Low	Moderate	Moderate	PLA, ABS, concrete	Construction, prototyping
SLA	Moderate	High	Moderate	Resins	Medical models
SLS	High	High	High	Nylon, metals	Aerospace, dental
Binder jetting	High	Moderate	High	Metals, ceramics	Industrial casting

Due of its affordability, FDM is widely used in developing nations, as evidenced by Ghana's adoption of RepRap machines in classrooms. The accessibility of FDM, where open-source platforms permit customization, is highlighted in the 2013 study (Mwanza & Mbohwa, 2018; OECD, 2021).

9.4 AM PROCESS OPTIMIZATION

Precise material and process control were essential to the 3D post office's success. The concrete mix struck a balance between inter-layer bonding (for structural integrity), flowability (for extrusion), and quick hardening (to support layers). This is consistent with the focus on material characterization as a crucial AM driver in the 2013 study (Calignano et al., 2017). AM technologies are changing globally to satisfy changing consumer demands. The University of Lagos in Nigeria produces educational models using FDM printers and PLA that is sourced locally. This lowers prices by 30%.

9.5 MATERIALS AND SUPPLY CHAIN CONSIDERATIONS

The majority of AM feedstock, which includes metals, resins, and thermoplastics, is imported. This raises costs and severely restricts access to materials in impoverished countries. Other AM uses still rely on imported materials, despite the Bengaluru post office using a specific concrete mix. Reflow's recycled polyethylene terephthalate (PET) filament in Pune, which lowers prices by 50%, and Rwanda's PLA manufacture from maize waste are two innovations that tackle this issue. This chapter discusses comparable programs in Latin America where inexpensive filament is produced through community-based recycling. Local supply connections are essential for AM to grow. The 2013 report emphasizes material development as a growth engine, with bio-based polymers like PLA providing sustainability. India is working to establish regional testing facilities to guarantee material quality under the direction of the Bureau of Indian Standards (BIS). This aligns with the chapter's suggestion to create regional supply networks in Table 9.4 (Horn & Harrysson, 2012).

TABLE 9.4
Material Challenges and Innovations

Challenge	Innovation	Impact
Import costs	Recycled PET filament, PLA from bio-waste	40–50% cost reduction
Testing facilities	Lack of local standards	BIS developing AM handbook
Supply chain	Community-based recycling initiatives	Supports circular economy

9.5.1 Material Innovations

Sustainable AM materials are the norm in emerging markets. By using 200 waste pickers to produce 1000 kg of PET filament a year in India, Reflow and other businesses support circular economies. Initiatives like Rwanda's PLA manufacture from agricultural waste cut Africa's reliance on imports by 40%, which is consistent with the chapter's emphasis on bio-based products. These initiatives tackle environmental issues since AM produces less waste than subtractive manufacturing, which, in certain techniques, can produce up to 90% junk.

The aerospace industry in India is investigating novel materials with higher conductivity and strength, like graphene-enhanced filaments. The adoption of AM is driven by material improvements, according to Guo and Leu (2013), with developing markets standing to gain from affordable, sustainable feedstocks. For these breakthroughs to be scaled, industry and university collaboration on research is essential, as demonstrated by the Indian Institute of Technology (IIT) Madras (Vaezi et al., 2012).

9.6 APPLICATIONS AND SECTORAL OPPORTUNITIES

- Building: AM's speed is demonstrated by the Bengaluru post office (43 days) and the 3D-printed school in Itanagar (two months). Similar to Mexico's housing projects mentioned later in the chapter, Tvasta Manufacturing in India constructed a 3D-printed home in 21 days. This can be compared to Uganda's prosthetic projects by e-NABLE.
- Healthcare: Divide by Zero's 3D-printed orthopedic casts and Anatomiz3D's SLA-based surgical models cut costs by 40–50%.
- Learning: Students are trained in AM at the University of Lagos in Nigeria, while the National Center for Additive Manufacturing (NCAM) in India improves science, technology, engineering and mathematics (STEM) skills.
- Farming: India's small and medium enterprises (SME) manufacture of agricultural implements is paralleled by Kenya's 3D-printed irrigation parts, which minimize downtime.
- Sustainable energy: India's AM-driven energy solutions, such as 3D-printed solar panel mounts, complement Bangladesh's solar lamp holders. Table 9.5 illustrates how AM is used.

TABLE 9.5
Additive Manufacturing Applications

Sector	Application	Impact
Construction	Post office, school, house	43 days, 2 months, 21 days construction times
Healthcare	Orthopedic casts, surgical models	40–50% cost savings
Education	STEM training, prototyping	Enhanced practical skills
Agriculture	Irrigation parts, tractor components	Improved equipment availability and reduced maintenance cost

9.7 SECTORAL APPLICATIONS

- AM has an impact that goes beyond major initiatives like the Bengaluru post office. AM provides essential healthcare services in emerging markets. In India, 50 hospitals employ Divide by Zero's 3D-printed orthopedic casts, which are lightweight and breathable, to treat 10,000 patients a year. The cardiac models from Anatomiz3D have facilitated more than 1000 procedures, while lowering expenses and enhancing results. The chapter's example from Uganda demonstrates comparable advantages; since 2020, e-NABLE has produced 500 prosthetic hands for children.
- AM lowers equipment downtime in remote agricultural areas. 3D-printed tractor parts are produced by Gujarati SMEs, which save repair periods by 60%. According to Kenya's Nakuru County trial, farmers can save 30% by adopting AM for irrigation connections. Applications for renewable energy include 3D-printed wind turbine parts from India, which boost the effectiveness of rural microgrids like Bangladesh's solar systems (Mueller, 2012).

9.8 CASE STUDIES AND SUCCESS STORIES

- India: The Itanagar school, the Bengaluru post office, and Tvasta's 3D-printed home serve as examples of AM's construction skills. Anatomiz3D's medical models facilitate surgical planning, which is consistent with the Indian instance discussed in this chapter.
- Kenya: 10,000 face shields were made by Nairobi's Gearbox FabLab during COVID-19, demonstrating AM's capacity for emergency response.
- South Africa: Use in healthcare is reflected in the 500 titanium implants that the Centre for Rapid Prototyping and Manufacturing (CRPM) provides each year.
- Brazil: As indicated in Table 9.6 of India's healthcare AM initiatives, Insper's prosthetic limb initiative in São Paulo has provided services to 1000 low-income patients.

TABLE 9.6
Applications and Socioeconomic Impact of Additive Manufacturing in Developing Countries

Country	Case study	Impact
India	Post office, school, house, medical models	Rapid infrastructure, 40–50% cost savings
Kenya	Face shields, ventilator parts	Emergency response capability
South Africa	Titanium implants	Advanced healthcare applications
Brazil	Prosthetic limbs	Improved mobility in underserved areas

9.8.1 India's Aerospace Sector

Startups such as 3D Incredible cut lead times from 12 weeks to two weeks by producing 3D-printed drone parts. This is in line with the South African example discussed in this chapter, where CRPM provides Boeing with titanium parts, generating 200 employment opportunities by 2024.

9.8.2 Nigeria's STEM Education

Similar to India's NCAM projects, the AM lab at the University of Lagos has improved mechanical engineering courses by creating 1000 educational models since 2022.

9.8.3 Mexico's Housing Program

Similar to the Tvasta programs in India, Apis Cor's 24-hour-built 3D-printed homes in Mexico help 10,000 households each year, who are experiencing housing shortages.

9.8.4 Rwanda's Circular Economy

Similar to India's Reflow project, local businesses use maize waste to manufacture 500 kg of PLA filament per month, helping 50 schools.

9.9 AM ADOPTION FACES CHALLENGES

- Infrastructure: As demonstrated in India's SME sector, access is restricted by 30% power disruptions and expensive equipment (USD 5,000–50,000).
- Skills shortage: NCAM addresses the issue of inadequate training by employing 1000 technicians a year.
- Material access: Costs are raised by import dependency, which is lessened by Rwanda's PLA and Reflow's filament.
- Standardization: Absence of regional guidelines, with BIS creating a guide to guarantee quality.

9.9.1 Challenges Analysis

Infrastructure continues to be a significant obstacle. The chapter's subject of intermittent electricity is shown in Table 9.7, which shows that 50% of SMEs in India report power interruptions that interfere with AM activities. Although there is limited scalability, since 2023, 100 SMEs have benefitted from pilot initiatives using solar-powered printers in rural India. Rwanda's local PLA production model, which produces one ton of filament per year for educational purposes, provides a model. The issue of standardization is worldwide. This is intended to be addressed by India's BIS handbook, which is anticipated to be released in 2026, but delays prevent its commercial use. Table 9.7 illustrates how the study by Guo and Leu (2013) places a strong emphasis on quality control, pointing out that flaws like residual stress might

TABLE 9.7
Key Challenges

Challenge	Description	Mitigation
Infrastructure	Power outages, high costs	Solar-powered printers, subsidies
Skills shortage	Limited AM training	NCAM training 1000 technicians
Material access	Import reliance	Recycled filament, PLA production
Standardization	No local standards	BIS handbook development

TABLE 9.8
Policy Recommendations

Policy	Action	Outcome
Subsidies	Fund AM equipment	Increased SME adoption
Regional hubs	University-industry collaborations	Innovation ecosystems
Standards	BIS handbook, ISO collaboration	Improved safety, scalability
Training	Expand to 5000 technicians by 2030	Reduced skills gap
Technology transfer	Partnerships with global AM firms	Enhanced capabilities

jeopardise structural integrity, which is an issue for 3D-printed buildings like the Bengaluru post office (Handfield et al., 2022).

9.10 ENABLING POLICIES AND ECOSYSTEM DEVELOPMENT

Policies for scaling AM are provided in Table 9.8.

- Financial aid: In 2024, India's NCAM will help 200 SMEs by providing equipment subsidies. University-industry partnerships, such as the AM Centre at IIT Madras, are examples of regional hubs that promote innovation.
- Standards: The recommendation for ISO collaboration presented in this study is aligned with the guidelines provided in the BIS handbook. Furthermore, the NCAM aims to train approximately 5,000 technicians in additive manufacturing technologies by 2030.
- Technology transfer: Collaborations with multinational companies such as COBOD improve AM capabilities, as indicated in Table 9.8 (Peprah et al., 2025; Yang et al., 2018).

9.10.1 Policy Framework

Government programs like "Make in India," which allots USD 10 million yearly for AM research and development, assist India's AM ecosystem. India's 50 AM innovation centers, which have been assisting 1000 entrepreneurs since 2022, are a reflection of the chapter's request for regional hubs. Two thousand entrepreneurs have been taught by the United Nations Industrial Development Organization (UNIDO)'s AM programs in Africa worldwide; India could follow suit. As this work suggests, tax breaks and low-interest loans are essential. Three hundred SMEs in India will benefit from a 2024 policy that provides 20% tax savings for AM equipment acquisitions. Guo and Leu (2013) highlighted government assistance for AM adoption and noted that limitations in intellectual property (IP) protection and certification impede expansion (see Table 9.8) (Allen et al., 2020; Joshi et al., 2012).

9.11 THE FUTURE OF AM IN DEVELOPING ECONOMIES

Sustainable materials, the Internet of Things (IoT), and artificial intelligence (AI) will influence AM's future. AI helps India's 3D Incredible cut down on print times by 15% and material waste by 20%. Sub-Saharan Africa's 21% CAGR prediction is comparable to India's 30% CAGR. As in Kenya, distributed manufacturing through local maker hubs is becoming more popular in India; by 2024, 100 community AM laboratories had been built.

Emerging technologies and AM are transforming each other. Real-time print quality monitoring using IoT-enabled printers in India has reduced errors by 25%. Brazil is an example of the chapter's concentration on hybrid AM subtractive systems, since aerospace companies there use 3D printing and computer numerical control (CNC) machining to increase precision by 30%. In 2024, 50 off-grid SMEs will be supported by solar-powered printers that are being tested in rural India to combat power shortages.

Using sustainable materials is important. Motivated by Rwanda, India plans to create 2 tons of PLA a year by 2026, resulting in a 40% cost reduction. The rise in AM is expected to be driven by material innovations, according to Guo and Leu (2013). This trend is seen in India's graphene-enhanced filaments for aerospace applications. With 200 hubs anticipated by 2028, distributed manufacturing models, backed by maker hubs, are expanding in India in line with the decentralized production goal, as illustrated in Table 9.9 (Allen et al., 2020; Horn & Harrysson, 2012).

TABLE 9.9
Future Trends

Trend	Description	Impact
AI/IoT integration	Smart sensors for print optimization	20% less waste, 15% faster printing
Recycled materials	PET filament, bio-based PLA	40–50% cost reduction, sustainability
Distributed models	Local maker hubs	Decentralized production
Hybrid systems	AM—subtractive integration	Enhanced precision, scalability

9.12 CONCLUSIONS

AM has the ability to propel sustainable industrialization, as seen by India's 3D-printed post office in Bengaluru, and wider uses in healthcare, education, and construction. To scale AM's influence in developing countries, and keep up with global trends in economic diversification and green manufacturing, cooperative policies, material advances, and training are essential.

AM is an effective tool for emerging economies because of its decentralized structure. India is positioned as an example for other markets due to its leadership, which is demonstrated by the Bengaluru post office and programs like NCAM. Developing countries can use AM to achieve self-reliant, inclusive growth by tackling issues through innovation, policy, and education. With India's AM ecosystem opening the door for worldwide influence, the objective of sustainable industrialization is achievable.

REFERENCES

Allen, A. J., Levin, I., & Witt, S. E. (2020). Materials research & measurement needs for ceramics additive manufacturing. *Journal of the American Ceramic Society*, 103(11), 6055–6069. https://doi.org/10.1111/jace.17369

Baumers, M., Beltrametti, L., Gasparre, A., & Hague, R. (2017). Informing additive manufacturing technology adoption: total cost and the impact of capacity utilisation. *International Journal of Production Research*, 55(23), 6957–6970. https://doi.org/10.1080/00207543.2017.1334978

Calignano, F., Manfredi, D., Ambrosio, E. P., Biamino, S., Lombardi, M., Atzeni, E., Salmi, A., Minetola, P., Iuliano, L., & Fino, P. (2017). Overview on additive manufacturing technologies. *Proceedings of the IEEE*, 105(4), 593–612. https://doi.org/10.1109/JPROC.2016.2625098 https://ieeexplore.ieee.org/abstract/document/7803596/

Guo, N., & Leu, M. C. (2013). Additive manufacturing: Technology, applications and research needs. *Frontiers of Mechanical Engineering*, 8(3), 215–243. https://doi.org/10.1007/s11465-013-0248-8

Haleem, A., Javaid, M., & Rab, S. (2020). Impact of additive manufacturing in different areas of Industry 4.0. *International Journal of Logistics Systems and Management*, 37(2), 239. https://doi.org/10.1504/ijlsm.2020.110578

Handfield, R. B., Aitken, J., Turner, N., Boehme, T., & Bozarth, C. (2022). Assessing adoption factors for additive manufacturing: Insights from case studies. *Logistics*, 6(2), 36. https://doi.org/10.3390/logistics6020036

Horn, T. J., & Harrysson, O. L. A. (2012). Overview of current additive manufacturing technologies and selected applications. *Science Progress*, 95(3), 255–282. https://doi.org/10.3184/003685012x13420984463047

Hou, Z. (2025). Research progress of volumetric additive Manufacturing. *Applied and Computational Engineering*, 186(1), 32–38. https://doi.org/10.54254/2755-2721/2025.gl27143

Joshi, P. C., Dehoff, R. R., Duty, C. E., Peter, W. H., Ott, R. D., Love, L. J., & Blue, C. A. (2012). Direct digital additive manufacturing technologies: Path towards hybrid integration. In *2012 Future of Instrumentation International Workshop (FIIW) Proceedings* (pp. 1–4). IEEE. https://doi.org/10.1109/FIIW.2012.6378353

Lanz, R., Miroudot, S., & Nordås, H. K. (2018). Emerging manufacturing technologies and developing countries (OECD Trade Policy Papers, No. 214).

Lydon, J., Silva, J., & Tallon, L. (2019). Additive manufacturing and development: What is the future? World Bank Working Paper Series. https://documents.worldbank.org

Mellor, S., Hao, L., & Zhang, D. (2014). Additive manufacturing: A framework for implementation. *International Journal of Production Economics*, 149, 194–201. https://doi.org/10.1016/j.ijpe.2013.07.008

Mueller, B. (2012). Additive manufacturing technologies—Rapid prototyping to direct digital manufacturing. *Assembly Automation*, 32(2), 140–141. https://doi.org/10.1108/AA.2012.03332BAA.010

Mwanza, B. G., & Mbohwa, C. (2018). Additive manufacturing and sustainability: An exploratory systematic review. *Procedia Manufacturing*, 21, 554–561. https://doi.org/10.1016/j.promfg.2018.02.153

OECD. (2021). *The digital transformation of SMEs*. OECD Publishing. https://doi.org/10.1787/bdb9256a-en

Peprah, J. A., Amoah, J., Kwarteng, K., Asante, D., Boateng, F., & Mensah, E. (2025). Artificial intelligence and additive manufacturing for resilient supply chain in Africa: A systematic literature review. *Future Business Journal*, 11, 54. https://doi.org/10.1186/s43093-025-00477-y

Rashid, A. A., Khan, S. A., Al-Ghamdi, S. G., & Koç, M. (2020). Additive manufacturing: Technology, applications, markets, and opportunities for the built environment. *Automation in Construction*, 118, 103268. https://doi.org/10.1016/j.autcon.2020.103268

Research and Markets. (2022)). *Global additive manufacturing market report 2022 to 2027—Featuring 3D Corporation, Optomec, Siemens and Renishaw among others*. Business Wire. www.businesswire.com/news/home/20221125005268/en/Global-Additive-Manufacturing-Market-Report-2022-to-2027---Featuring-3D-Corporation-Optomec-Siemens-and-Renishaw-Among-Others---ResearchAndMarkets.com

Tegethoff, T., Santa, R., Bucheli, J. M., Cabrera, B., & Scavarda, A. (2025). Navigating industry 4.0: Leveraging additive technologies for competitive advantage in Colombian aerospace and manufacturing industries. *PLoS ONE*, 20(2), e0318339. https://doi.org/10.1371/journal.pone.0318339

Vaezi, M., Seitz, H., & Yang, S. (2012). A review on 3D micro-additive manufacturing technologies. *The International Journal of Advanced Manufacturing Technology*, 67(5–8), 1721–1754. https://doi.org/10.1007/s00170-012-4605-2

Yang, Y., Song, X., Li, X., Chen, Z., Zhou, C., Zhou, Q., & Chen, Y. (2018). Recent progress in biomimetic additive manufacturing technology: from materials to functional structures. *Advanced Materials*, 30(36), 1706539, 1–34. https://doi.org/10.1002/adma.201706539.

10 Thermo-Mechanical Behavior of Additively Manufactured Materials: Insights and Analysis

Mufaddal Huzefa Shakir, Akant Kumar Singh, Anant Krishan Pun, Naveen Mani Tripathi, and Arun Gupta

10.1 INTRODUCTION

Additive manufacturing (AM) is a rapidly expanding manufacturing technique, which is utilized to develop three-dimensional (3D) parts or components. The AM method consists of building different components or parts one layer after another, utilizing a 3D model obtained from the computer (Parupelli & Desai, 2019). 3D printing can use various materials, like composites, ceramics, metals, and polymers, to create the 3D printed part, depending on the requirements and application. This technique can easily build complex shapes and geometries. There are different AM techniques, which can be categorized on the basis of the state of raw materials used, like solid-based systems, powder-based systems, and liquid-based systems (Zhou et al., 2024). Figure 10.1 provides a detailed illustration of the various AM processes. AM has several benefits over traditional and subtractive manufacturing techniques. Some of these involve design flexibility, mass customization, waste reduction, and ease of creating complex geometry. Owing to the increasing demand of this technology, 3D printing has extended its usage to different industries, like healthcare, the automobile industry, electronics, aerospace, sports, and construction. 3D printing is now frequently used to make complex lightweight structures (Fu et al., 2022; Jadhav & Jadhav, 2022).

Different AM techniques usually utilize a combination of the following steps:

- Analyzing and designing the 3D CAD model
- Changing the 3D CAD model into STL file format
- Passing the STL file to the 3D printing machine
- Making the equipment ready

DOI: 10.1201/9781003643531-10

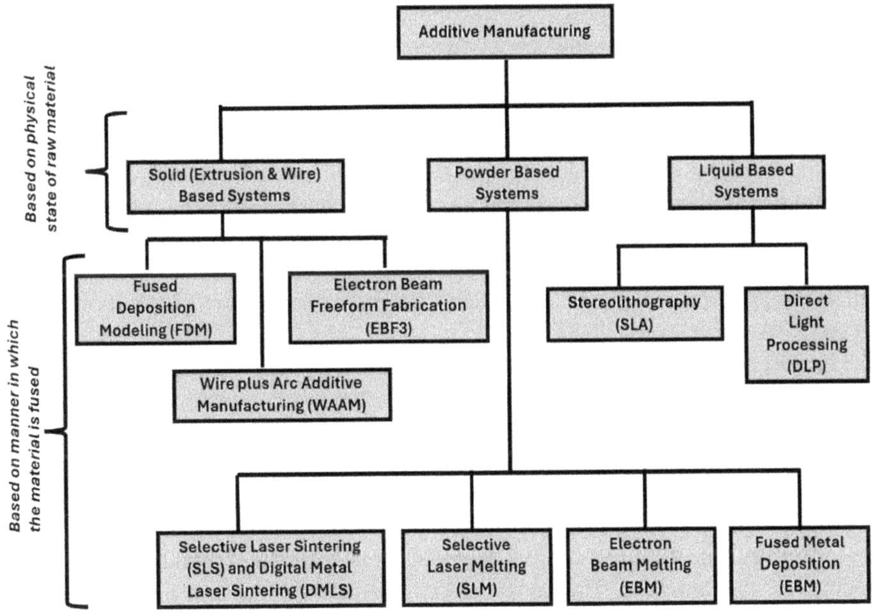

FIGURE 10.1 Categorization of different 3D printing processes.

- Creating the part or component
- Withdrawal of the built part or component
- Post-processing of the part or component
- Usage for the desired application

Additive manufacturing of polymer matrix composites (PMCs) has opened new opportunities in the design of low-weight structures and the development of functional components. In PMCs, thermosetting polymers, such as polyester, epoxy, and phenol formaldehyde, or thermoplastic polymers, such as nylon, polycarbonate, polypropylene (PP), or polyethylene, reinforced with synthetic or natural fibers such as carbon, glass fiber, kevlar, jute, flax, or hemp, are usually used to fabricate the composites. Additive manufacturing gives precise control over fiber orientation and resin distribution, paving the way to create highly customized components. Fused deposition modeling (FDM) and stereolithography (SLA) are the most utilized AM methods to develop fiber-reinforced composites, enhancing their mechanical and thermal behavior (Parandoush & Lin, 2017). Employing AM methods in the creation of metal matrix composites (MMCs) has advanced AM's applicability in high-performance engineering environments. MMCs are usually composed of a metallic matrix like magnesium, copper, nickel-based alloys, aluminum, or titanium, reinforced with ceramic or metal particles to enhance strength, thermal stability, hardness, and wear resistance. Selective laser melting (SLM) and direct metal laser sintering (DMLS) are 3D-printing techniques that are generally utilized to create MMCs with complex geometries and controlled microstructures. These methods allow for the accurate addition of reinforcements within the metal matrix, overcoming challenges like aggregation and poor bonding (Murr, 2018). The AM of ceramic matrix composites (CMCs) demonstrates an important step in developing materials that can survive high temperatures and harsh environments. CMCs are usually composed of silicon carbide particles or fibers as reinforcement, enclosed in a ceramic matrix, providing superior mechanical behavior and thermal resistance at high temperatures. The traditional manufacturing of CMCs is quite expensive and time-intensive due to high sintering temperatures and complex processing requirements. Nonetheless, 3D-printing techniques like SLA-based ceramic printing or robotic material extrusion enable the creation of intricate and porous ceramic structures with high-dimensional accuracy (Aramian et al., 2020).

The properties of 3D-printed materials are generally impacted by various printing factors like print or scan speed, thickness of the layer, nozzle and printing bed temperatures, layer height, fill density, infill pattern, and nozzle diameter (Prabhakar et al., 2021). Printing or scan speed impacts both the quality of adhesion between the layers and the overall surface finish of the part. High printing speed can lead to incomplete fusion, resulting in poor interlayer bonding and decreased part strength. Low printing speed offers better material deposition and bonding, improving mechanical performance. The layer thickness of AM parts plays a significant role in deciding on better surface resolution and mechanical behavior. Thinner layers give a smoother surface finish, along with improved strength due to better layer infusion, while thicker layers may result in reduced mechanical uniformity. Larger layer heights also reduce the build time, but weaken interlayer bonding (Ingrassia et al., 2016). The nozzle temperature of the 3D printer plays a vital role in proper material flow and bonding. High temperatures may lead to the degradation of the material or induce thermal stress, while lower temperatures may cause poor extrusion and weak bonding. A higher printing bed temperature improves first-layer adhesion, reducing the phenomenon of warping. For 3D-printed composites, an optimal bed temperature ensures uniform melt flow and better wetting of fibers, improving bonding. Furthermore, higher infill density in 3D-printed composites leads to higher load-bearing capacity, less deformation, and good thermal conductivity, whereas, for lower infill density, the weight of the material is decreased, but the strength of the material might be affected. With respect to the nozzle diameter, a small diameter (0.2 to 0.4 mm) is not desirable for long fiber composites, but is better for a pure matrix. A small diameter provides a better surface finish with good interlayer bonding. A medium diameter (0.4 to 0.6 mm) provides a good trade-off between detail and speed. A large diameter (0.6 to 1.0 mm) allows for the faster deposition of materials and enables long fibers to easily pass through it. Considering these print parameters is very important in getting the required material properties, dimensional accuracy, and reliability in AM-printed components (Mostafaei et al., 2021; Mushtaq et al., 2023).

This chapter aims to provide insights into the thermal and mechanical behavior of AM materials such as PMCs, MMCs, and CMCs. The chapter addresses various studies conducted by researchers on these AM materials. Understanding the thermo-mechanical properties of these materials is crucial in determining their application in several industrial sectors. Hence, the discussions on 3D-printed materials in this chapter are beneficial to researchers who are advancing in the area of 3D printing.

10.2 THERMO-MECHANICAL BEHAVIOR OF PMCS

10.2.1 Mechanical Behavior of PMCs

FDM is a frequently utilized process for developing PMCs. 3D-printed PMCs are lower in cost, require less energy, enable faster prototyping, and are lightweight compared to MMCs and CMCs (Jaganathan et al., 2024). A study utilized FDM to examine the properties of acrylonitrile butadiene styrene (ABS)/TiO_2 composite samples. The research assessed the tensile properties of these composite materials. The composite samples to be tested were additively manufactured in two different build orientations (horizontal and vertical), and the properties between the composite samples compared. The TiO_2 was kept at 5 wt.% in the ABS matrix to create the composites. The tensile strength of the horizontally and vertically printed composite samples was discovered to be 32.9 and 16.7 MPa, respectively. Additionally, the breaking elongation values were discovered to be 3.77 and 1.61% for the horizontally and vertically printed composites, respectively. Only a slight difference was observed in the mechanical behavior of ABS/TiO_2 composite specimens in comparison to neat ABS specimens (Torrado et al., 2015). The mechanical strength of additively manufactured ABS composite samples is presented in Figure 10.2.

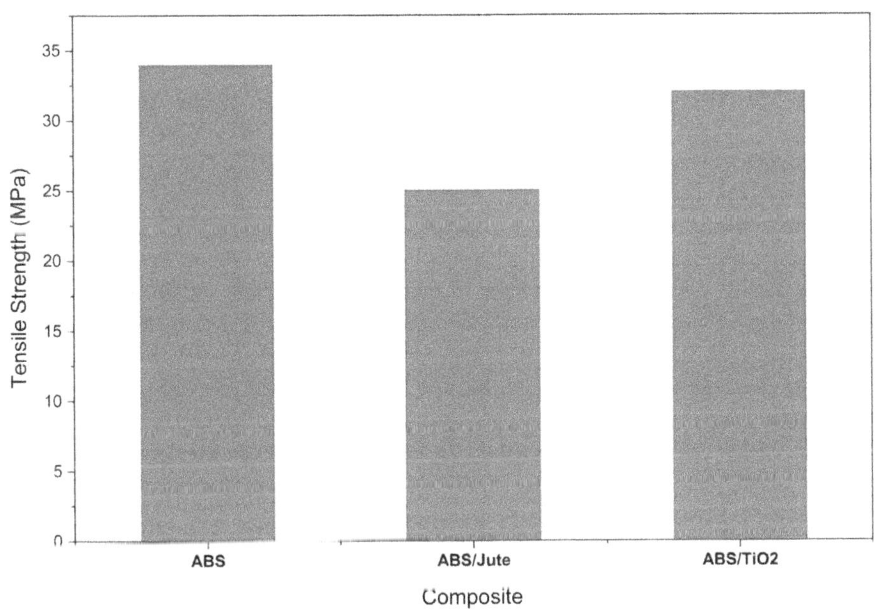

FIGURE 10.2 Mechanical strength of ABS/Jute, ABS/TiO_2 and neat ABS composite samples with reinforcement at 5 wt.%.

A research study developed short carbon fiber-reinforced ABS composite samples using FDM. The properties of the FDM-printed composite specimens were compared with the compression-molded specimens. The reinforcement loading used was in the range of 10–40 wt.%, with the fiber size being 0.2–0.4 mm in length. The temperature of the nozzle and printer was kept at 205 and 85°C. The height layer was kept at 0.2 mm. The tensile properties of the AM composite specimens were observed to increase with a rise in the fiber content. It was also noted that 3D-printed composite specimens had a high fiber orientation, whereas compression-molded samples had a lower fiber orientation. However, the mechanical behavior of the composite specimens created by the compression molding process was comparable to AM composite samples (Tekinalp et al., 2014). Another study involved the fabrication of glass fiber/polyamide 12 (PA12) composite specimens. This research utilized the multi-jet fusion (MJF) 3D printing technique to manufacture the composites. MJF-printed composite materials usually have lower mechanical properties, even with the incorporation of glass fibers. Therefore, a high-temperature annealing method (173°C for five hours) was introduced to enhance the mechanical behavior of the glass fiber/PA12 and pure PA12 composite specimens. The tensile strength and elastic modulus of neat PA12 composite samples were observed to be 56.17 and 2042 MPa, respectively, whereas for glass fiber-reinforced PA12 composite samples, it was discovered to be 59.35 MPa and 5.25 GPa, respectively. It was simultaneously noted that the properties of MJF AM GF/PA12 composites were higher than selective laser sintering (SLS)-printed composites, as discovered in earlier studies. It was demonstrated that a high-temperature annealing process had greatly influenced the mechanical behavior of GF/PA12 composite samples (Liu et al., 2021).

Another study analyzed the impact of different reinforcements on the tensile properties of AM ABS composite materials. Reinforcements like jute fibers and TiO_2 were kept at 5 wt.% in the ABS polymer. Furthermore, a polymeric blend of ABS was developed with 5 wt.% of thermoplastic elastomers (TPE). The composites were created by FDM. The layer height of the material was kept at 0.27 mm, and the nozzle temperature was around 230°C. The mechanical and fracture surface characteristics of composite specimens were studied and compared. The tensile strength of the ABS/TiO_2 composite specimen was greater in contrast to ABS/jute, ABS/TPE, and neat ABS composite specimens. The fracture surface analysis revealed that, instead of expecting ductile fracture, as seen in the ABS/jute composite, the fracture characteristic of the ABS/TiO_2 composite showed brittle fracture (Torrado et al., 2014). Figure 10.3 depicts the mechanical strength of carbon fiber composites created by FDM. In another study, lyocell/polylactic acid (PLA) composites were fabricated utilizing FDM. A PLA matrix was modified with maleic anhydride, while lyocell fibers were fibrillated prior to the melt compounding process. The fiber weight percentage was varied from 10 to 30 wt.%. The mechanical behavior of the composites was assessed. It was found that composites with 30 wt.% fibers had enhanced tensile properties in comparison to composite samples with unmodified fibers and matrix. The post-processing annealing treatment of the composite samples also contributed to an improvement in the tensile properties (Gauss et al., 2023).

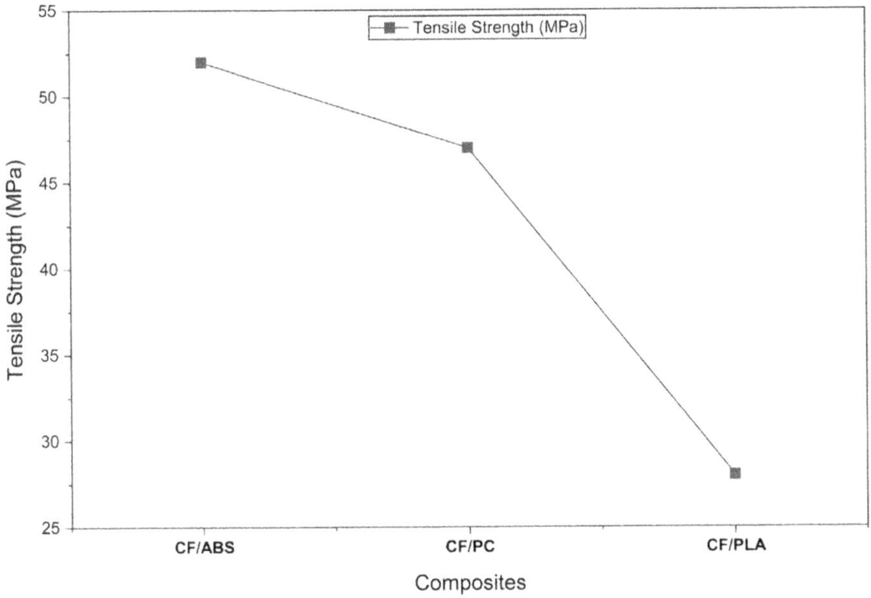

FIGURE 10.3 Mechanical strength of AM carbon fiber thermoplastic composites.

In another study, metal/polymer composite filaments were created for the 3D-printing FDM process. The matrix material used was ABS, and reinforcement was in the form of metal powders such as Cu and Fe. The reinforcement loading was varied from 10 to 30 wt.%, respectively. Two types of filaments were developed—ABS/Cu and ABS/Fe—and the properties of these materials were studied. The tensile strength of the pure ABS composite specimen was noted to be 45.7 MPa. After the addition of Cu powder from 10 to 30 wt.%, it was discovered that the tensile strength of the composite samples decreased from 45.7 to 26.5 MPa. A similar trend was seen in the case of Fe powder. The tensile strength of the ABS/Fe composite samples was observed to decrease with an increase in the Fe powder's weight percentage. The study showed that the incorporation of Cu and Fe particles in the ABS matrix did not enhance its tensile strength (Hwang et al., 2015). Another research study analyzed the mechanical properties of ABS/boron nitride (BN) MMCs, which were fabricated by both 3D printing and injection molding. A twin screw and single screw extruder was used to create the filaments for the injection molding and FDM 3D-printing process. The BN content was varied from 5 to 35 wt.%, whereas the infill orientation of the AM composite samples was varied to study the impact on its properties. For the AM process, the nozzle and print bed temperatures were around 240 and 80°C, the layer height was maintained at 0.2 mm, and the printing speed was 50 mm/s with a 100% fill density. The composite specimens were printed in both 0–90° and 45–45° infill orientations. The impact strength and bending properties of the composite specimens were analyzed. With an increase in BN content, the impact strength of the ABS/BN composite samples was found to decrease. At lower BN loadings, the impact strength of the injection-molded ABS/BN composite samples was found to be greater in comparison to the AM composite specimens. With respect to infill orientation, no big difference was observed in the impact strength of the composites. At 25 wt.% BN loading, the bending modulus of AM composites was observed to be 3.93 GPa. The infill orientation only had a slight impact on the bending properties of the composites, where a 45–45° pattern showed higher flexural properties in comparison to a 0–90° infill pattern (Quill et al., 2018).

To enhance the mechanical behavior of AM composite specimens, carbon fiber (CF) was co-extruded with a thermoplastic matrix by a 3D printing process known as towpreg co-extrusion. By utilizing this process, continuous CF coated with epoxy matrix and polyamide as the thermoplastic matrix was co-extruded, thus obtaining an AM two-matrix composite. The continuous CF epoxy filament composite comprised 1.5-K CFs in a tow. By observing the influence of moisture content, the mechanical behavior of the composite materials was assessed. The drying treatment was carried out by keeping the specimens in a hot air oven at 70°C for two hours. A reduction in 0.56% by weight was observed for the specimens after the drying process. The drying process had a greater impact on the mechanical behavior of the composite samples. The bending and tensile strength of the composite materials were discovered to be 151.05 and 392.78 MPa, respectively. A rise in the bending and tensile modulus was also seen for the composite specimens (Andreozzi et al., 2024). The properties of several PMCs are reported in Table 10.1.

10.2.2 Thermal Behavior of PMCs

The thermal behavior of PMCs is an important aspect that determines their performance under varying temperature conditions, mainly in the aerospace, automotive, and electronics industries. PMCs generally have low thermal conductivity in comparison to MMCs and CMCs. Thermosetting polymers typically have better thermal stability in comparison to thermoplastic polymers. The utilization of synthetic or natural fibers and fillers improves the overall thermal behavior of PMCs. The AM technique and post-processing significantly impact the thermal behavior of the materials. It is to be noted that PMCs require careful temperature control during 3D printing to avoid deformation (Bute et al., 2023). A research study explored the thermal behavior of a novel 3D-printed metal-based PLA composite. Different composites were created using various metal powders like bronze, copper, magnetic iron, and stainless steel as reinforcement in the PLA matrix. The thermal conductivity of AM metal/PLA composite materials was analyzed. The thermal conductivity of bronze/PLA composites was discovered to be greater than copper/PLA composites by 21.92%, even though the thermal conductivity of pure copper and bronze were 375 and 45 W/m K, respectively. On other hand, magnetic iron/PLA composites had better thermal conductivity than stainless steel/PLA composites. It should be noted that the thermal conductivity of these composites was superior to neat PLA (Laureto et al., 2017).

Dynamic mechanical analysis (DMA) is an important method that is used to measure the thermo-mechanical behavior of material, giving insights about its glass transition temperature (T_g), viscosity, storage, loss, and complex modulus (Saba

TABLE 10.1
Mechanical Properties of PMCs

Material	Matrix/fiber content	Method	Printing speed (mm/s)	Layer height (mm)	Tensile strength (MPa)	Bending strength (MPa)	Breaking elongation (%)	Reference
Jute/ABS	5 wt.% jute fibers	FDM	40	0.27	24.25	-	4.25	Torrado et al., 2015
Carbon fiber/ABS	40 wt.% fibers	FDM	-	0.2	65.0	-	-	Tekinalp et al., 2014
Glass fiber/polyamide 12	20 wt.% glass fibers	Multi-jet fusion	200	0.08	59.35	-	4.81	Liu et al., 2021
TiO_2/ABS	5 wt.% TiO_2	FDM	55	0.27	32.2	23.8	-	Torrado Perez et al., 2014
Lyocell/PLA	30 wt.% lyocell fibers	FDM	30	0.1	58.4	-	3.15	Gauss et al., 2023
Glass fiber/polycarbonate	5 wt.% glass fibers	FDM	45	0.3	28.0	-	-	Tikhani et al., 2024
Carbon fiber/polycarbonate	10 wt.% carbon fibers	FDM	75	0.2	47.0	94.0	-	Gupta et al., 2020
Sisal/gypsum powder	1.5 % sisal fibers by volume	Binder jetting	-	0.1	-	12.52	-	Coelho et al., 2019
Carbon fiber/PLA	15 wt.% carbon fibers	FDM	60	0.2	28..0	-	2.1	Ogaili et al., 2024
Kevlar/nylon	8 wt % fibers	FDM	-	0.1	88.18	-	4.31	Shi et al., 2021
Ag-TiO_2/SLR	1 wt % Ag-TiO_2 nano-particles	SLA	-	-	44.7	70.7	3.6	Mubarak et al., 2020

(continued)

TABLE 10.1 (Continued)
Mechanical Properties of PMCs

Material	Matrix/fiber content	Method	Printing speed (mm/s)	Layer height (mm)	Tensile strength (MPa)	Bending strength (MPa)	Breaking elongation (%)	Reference
Iron/ABS	10 wt.% iron particles	FDM	-	-	43.4	-	6.1	Hwang et al., 2015
Boron nitride/ ABS	35 wt.% BN particles	FDM	-	-	-	38.0	-	Quill et al., 2018
Aramid fiber/PLA	20 wt.%	Nozzle impregnation	10.0	0.6	300.0	91.0	-	Kilinc et al., 2025
Betel nutshell fiber/ PLA	2.5 wt.%	FDM	-	-	22.0	33.22	-	Arora et al., 2025
Short carbon fiber/ nylon	35 wt.%	FDM	100	-	91.45	156.82	-	Wang et al., 2025
Basalt fiber/PLA	25 layers	FDM	-	-	175.0	-	3.2	Jung et al., 2025
Wood flour/ polypropylene	20 wt.% wood flour	FDM	20	1.5	10.34	17.54	-	Hwang et al., 2024

et al., 2016). A study analyzed the thermo-mechanical behavior of pure ABS and an ABS blend containing ultra-high molecular weight polyethylene (UHMWPE) and styrene ethylene butadiene styrene (SEBS). The blend of ABS/UHMWPE/SEBS was in the ratio of 75:25:10. The materials were evaluated at a 100 Hz frequency using a TA instrument. At 40°C, the complex viscosity of pure ABS and the ABS blend was found to be 80.21 and 27.92 MPa, respectively. It was to be noted that, at 100°C, the ABS blend was more reactive to an increase in temperature than the neat ABS. The complex viscosity was discovered to be 67.47 and 21.49 MPa, respectively, for the pristine ABS and the ABS blend. Moreover, the T_g of the pristine ABS and the ABS blend was found to be 112.62 and 115.47°C, respectively. The DMA results also revealed the ABS blend to be less viscous than pure ABS above 140°C (Torrado et al., 2015). Another research study used the DMA technique to study the properties of AM neat PLA and lyocell/PLA composite specimens. The composite specimens were examined in the temperature range of 25–140°C. The lyocell fiber composition for these composites was kept at 30 wt.%. For lyocell/PLA composites, the PLA used was modified with maleic anhydride, and the reinforcement went through the process of fiber fibrillation. After the fabrication, the composites had gone through the process of heat treatment at 105°C for two hours in a laboratory oven. At around 60°C, there was a drop in the storage modulus (E`) for all compositions, which depicted a glassy to rubber transition. At high temperatures, the E` values were observed to be lower, but the difference in E` between pure PLA and PLA composites was more pronounced, meaning that lyocell/PLA composites had a higher storage modulus than neat PLA. The introduction of lyocell fibers and increasing the crystallinity of PLA by heat treatment greatly enhanced the properties of the composite materials, increasing the E` value by 38% at 60°C and 200% at 80°C in comparison to pure PLA (Gauss et al., 2023). Table 10.2 displays the DMA results of various composite materials.

Thermogravimetric analysis (TGA) and differential scanning calorimetry (DSC) are two commonly utilized thermal analysis techniques used to characterize materials. TGA analyzes the weight loss of material as it is heated or cooled, whereas DSC

TABLE 10.2
Storage Modulus Results of Composite Materials at Different Temperatures (Gauss et al., 2023; Torrado Perez et al., 2015)

		Storage modulus (E`) MPa				
Material		**At 25°C**	**At 40°C**	**At 60°C**	**At 80°C**	**At 100°C**
Neat ABS	-	1511.00	-	-	1349.00	
Neat PLA	3214.00		96.62	7.68	-	
PLA-MA	3053.00	-	16.23	1.29	-	
ABS/UHMWPE/SEBS	-	530.40	-	-	436.10	
Lyocell/PLA—untreated	5099.00	-	907.7	63.13	-	
Lyocell/PLA—heat treated	5347.00	-	3461.00	14924.00	-	

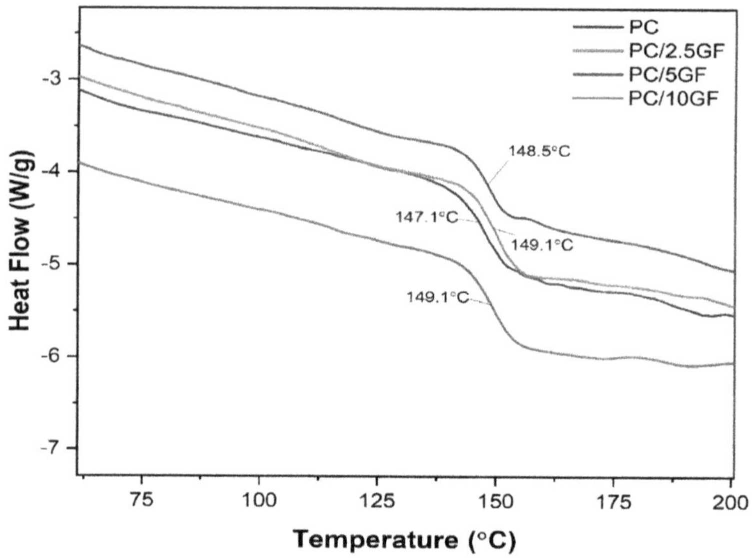

FIGURE 10.4 DSC curves of glass fiber/PC composites (Tikhani et al., 2024). Reproduced with permission from Elsevier.

examines the heat flow in the material associated with transitions such as melting, glass transition, or chemical reactions. These techniques help to optimize the 3D-printing process and assess material properties for different applications (Nurazzi et al., 2021). A study carried out the thermal analysis of AM E-glass fiber-reinforced polycarbonate (PC) composites. For DSC analysis, samples of 10 mg were used. The temperature range was 20–350°C at a heating rate of 10°C per minute under nitrogen atmosphere. For the TGA measurement, samples of 10–20 mg were used. The DSC curves showed that the T_g of the composites was shifted to higher temperatures due to the incorporation of glass fibers, which indicated an improved thermal behavior of composite samples in comparison to pure PC (Figure 10.4). It is well known that reinforcements can hinder the molecular mobility of polymers. The TGA results depicted that the composites were thermally stable up to 470°C. It was observed that the inclusion of 10 wt.% fiberglass in PC resin increased the degradation temperature of composites somewhat to 477°C in comparison to neat PC, which had a degradation temperature of 470°C. The good interaction between PC and glass fibers could be attributed to improved structural integrity and thermal behavior (Tikhani et al., 2024). Another research study investigated the thermal behavior of AM sisal fiber-reinforced gypsum powder composites. The sample was heated at 10°C per minute under the influence of a nitrogen atmosphere until 900°C. By comparing the DSC curves of neat gypsum powder and sisal/gypsum composites, it was not possible to notice the degradation of the main components of the sisal fibers, such as cellulose, hemicellulose, or lignin. Thus, non-observance of the decomposition of natural fibers in the AM specimens could be clarified by the presence of low sisal fiber content (1.5% by volume fraction). Sisal fibers could not be concluded to have an impact on the thermal stability of the composite materials. The difference in weight loss, as depicted by

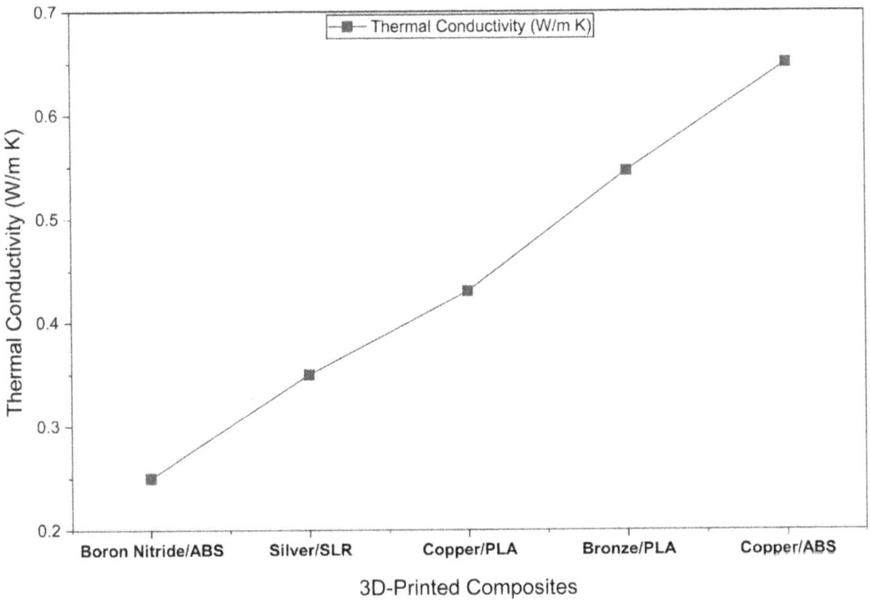

FIGURE 10.5 Thermal conductivity of PMCs.

the TGA curves, between the sisal-gypsum composites and the neat gypsum powder was found to be 2.14%. Therefore, it could be presumed that the distribution of sisal fibers within the gypsum powder bed and in 3D-printed specimens was homogeneous (Coelho et al., 2019). Another research study examined the thermal behavior of novel epoxy-coated carbon fiber-reinforced polyamide 12 (PA12) composites, which were manufactured utilizing the 3D-printing method. The carbon fiber weight fraction was maintained at 60 wt.%. For thermal analysis, samples of about 10 mg were utilized under the influence of a nitrogen atmosphere, with a heating time of 10°C per minute. The thermal analysis of the composites was carried out under two different conditions (dried and undried). Under the first condition, the composites were stored at an ambient temperature (undried), and under the second condition, the composites were desiccated in a hot air oven at 80°C for two hours (dried). The differential thermogravimetric analysis (DTG) and TGA curves showed similar thermal degradation behavior. For the undried composite specimen, a weight loss of 0.50% was observed before 130°C, confirming the presence of moisture content in the composite specimen, which impacts its mechanical performance. For the PA12, the main peak was observed at 454°C in the DTG curves as mass loss due to the thermal degradation of the PA12. At 384°C, a mass loss could also be noted for the undried composite specimen. This value was somewhat lower than the DTG peak of the standalone epoxy matrix in the carbon fiber bundle at 415°C, indicating that contact with the PA resin complicates the resin's breakdown process. At a temperature of 900°C, the final residual char of polyamide matrix was observed at 1.6%, whereas it was 45.8 and 46.6% for the dried and undried composite specimens, respectively (Andreozzi et al., 2024). Figure 10.5 demonstrates the thermal conductivity of PMCs.

10.3 THERMO-MECHANICAL BEHAVIOR OF METAL MATRIX COMPOSITES

10.3.1 Mechanical Behavior of MMCs

The properties of MMCs are influenced by both the matrix and the reinforcement used. Generally, 3D-printed MMCs display a high strength-to-weight ratio, hardness, and wear resistance when ceramic particles such as SiC, Al or TiO_2 are evenly distributed within the metal matrix. 3D printing methods like SLM or direct energy deposition (DED) create rapid thermal cycles, which can result in residual stress, microcracks, or porosity. This may lower the mechanical performance of AM MMCs if not properly controlled. Thus, post-processing treatments like hot isostatic pressing (HIP) or annealing are necessary to enhance densification and reduce internal defects. MMCs created with a well-optimized 3D-printed process display high mechanical performance, enabling them to be utilized in applications in the aviation, automobile, and defense sectors (Dadkhah et al., 2021). Studies have been performed where the mechanical behavior of AM MMCs have been investigated. One study analyzed the mechanical performance of fiber metal laminates (FMLs) based on AM composite materials. Here, FDM was utilized to create continuous carbon fiber-reinforced onyx composite materials using a Mark Two 3D printer by Markforged. These composites were positioned between the layers of aluminum alloy to create FMLs. Different stacking configurations of FMLs were developed, such as FML-1, which comprised Al/composite/Al, and FML-2 consisting of an Al/composite/Al/composite/Al-type arrangement. The composite volume fraction was kept at 43 and 50% for FML-1 and FML-2, respectively. To achieve bonding between the metal and the composite, a modified polyethylene film was used between the metal-composite layers. The mechanical behavior of FMLs was investigated. The experimental outcomes displayed that the tensile strength of FMLs was found to be between pure aluminum alloy and a composite. In terms of impact strength, FMLs were observed to exhibit improved low- and high-velocity impact strength in comparison to other compositions (Yelamanchi et al., 2020).

A new technique for the 3D printing of MMCs by the thermal decomposition of salts (TDS) was investigated in another study. In this process, inorganic salt was printed to develop MMCs. This was followed by bulk sintering. In the time of sintering, the 3D-printed inorganic salt decomposed into fine ceramic particles to create MMCs. The inorganic salt used was aluminum sulfate, which then decomposes into aluminum oxide particles during the sintering process. Bronze metal was used as the matrix material. The layer thickness was varied from 0.1 to 0.5 mm, depending on the ceramic agent and metal powder. The sintering process took place in an inert environment with argon gas. The impact of different sintering parameters like temperature and time period on the mechanical behavior of bronze/alumina composites was studied. There was a noticeable enhancement in the yield strength of the composite materials with a rise in the sintering temperature. At a temperature of 770°C and increasing the sintering time span from one to two hours, a rise in the yield strength of the composite was observed. For the same parameters, there was no significant enhancement in yield strength when the time was increased from zero to one hour (Yoozbashizadeh et al., 2018). Another research study investigated the mechanical behavior of AM aluminum alloy (AlSi10Mg)-reinforced SiC metal-matrix composites. The size of the SiC particles was in the range of 10–30 microns. The properties were analyzed by varying the composition of the aluminum alloy and SiC particles. The composite samples also underwent HIP heat treatment after fabrication, and were compared with untreated samples. It was discovered that untreated composites with 2 wt.% SiC reinforcement and 0° orientation had better mechanical properties in comparison to heat-treated composites. The composites with 5 wt.% SiC reinforcement had lower mechanical properties compared to 2 wt.% SiC composites (Bai & Liu, 2024). Similarly, a research study examined the influence of solution-aging and annealing heat treatment on the mechanical behavior of cold-sprayed AM aluminum alloy (AA2024)/aluminum oxide composites. The size of both ceramic and metal powders was in the range of 15–40 microns. Prior to 3D printing, the metal and ceramic powders were mixed in a ball milling machine in the ratio of 3:1. For heat treatment, different parameters like temperature and time were used to study their effects on the properties of the composites. It could be seen that, irrespective of the heat treatment parameters, the hardness of the heat-treated composites decreased in comparison to the untreated composites. The results showed that solution-aging-treated composites at 500°C and six hours provided the highest value of tensile strength and breaking elongation. It was concluded that solution-aging treatment was better than annealing to achieve a greater mechanical behavior of the composite samples (Wu et al., 2022). The tensile strength of different MMCs is shown in Figure 10.6.

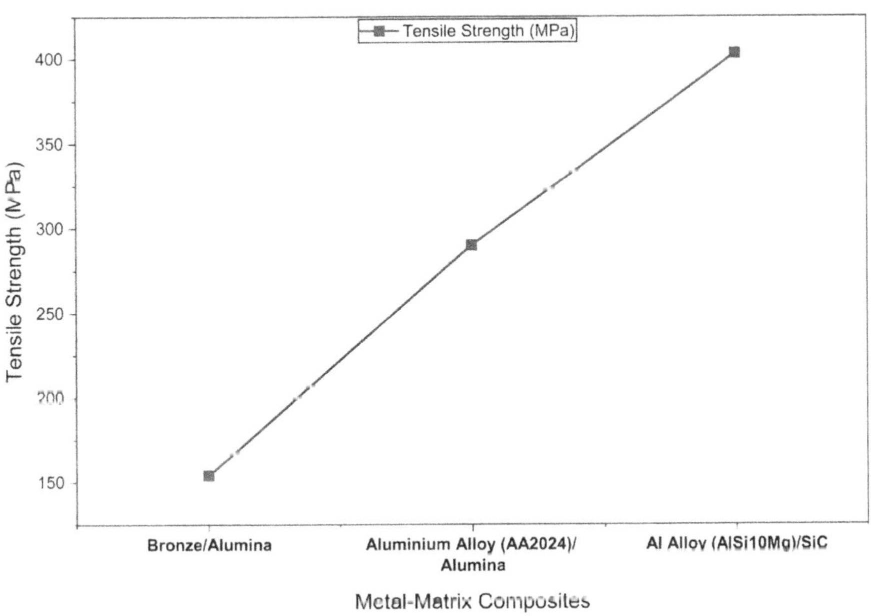

FIGURE 10.6 Tensile strength of different MMCs.

For high-temperature applications, a research study focused on fabricating a 3D-printed Inconel 625 (Inc625) alloy-based MMCs. Different ceramic particles, such as TiC, ZrC, TiB_2, and ZrB_2, were utilized as reinforcement in creating MMCs. Laser powder bed fusion (LPBF) was employed as a 3D printing technique to fabricate these composites. The average size of the ceramic powders was around 2 μm. For each composite, the reinforcement content of the ceramic powders was kept at 2 wt.%. The mechanical performance of these MMCs was analyzed at higher temperatures. The tensile strength of composites was evaluated at a temperature of 800°C, whereas creep resistance was investigated at a temperature of 800°C and a tensile stress of 150 MPa. It was observed that an Inc625/TiB_2 composite exhibited the highest tensile strength, but had a lower ductility. Among these, Inc625/ZrC showed lower strength-ductility performance. With respect to creep strain, the Inc625/ZrB_2 composite showed the highest creep stain at 57.2%, while baseline Inc625 showed the lowest creep strain at 8%. On the other hand, Inc625/TiB demonstrated the longest duration of creep rupture time at 547 hours, whereas Inc625 showed the lowest duration at 32 hours. It could be concluded that Inc625/TiB_2 composites display remarkable mechanical behavior for high-temperature applications (Tekoğlu et al., 2024). Another research study utilized the LPBF 3D-printing method to create Inconel 718 (Inc718) alloy-based MMCs. Ceramic particles such as zirconium diboride (ZrB_2) were used as reinforcement in the Inc718 alloy. The diameter of the ceramic particles was less than 100 nm, and reinforcement loading was kept at 2 wt.%. The mechanical performance of these MMCs was analyzed at an elevated temperature of about 800°C. The scanning electron microscopy (SEM) images showed that the 3D-printed Inc718/ZrB_2 composite had a lower porosity and smaller grain particle size compared to an AM Inc718 alloy under similar printing conditions. At 650°C, the tensile testing results showed that an Inc718/ZrB_2 composite had a 15% increase in tensile and yield strength compared to a pure Inc 718 alloy, although a significant loss in ductility was observed (5%). Furthermore, the Inc718/ZrB_2 composite demonstrated a rise in ductility at 800°C, attaining around 10% breaking elongation, while simultaneously maintaining higher mechanical strength in comparison to pristine Inc718 alloy (Tekoğlu et al., 2024). The mechanical strength of the Inconel alloy 625 and 718, along with their composites, is reflected in Figure 10.7.

Research explored the mechanical behavior of titanium alloy metal matrix composites, which were fabricated by DED. The reinforcement used in this study was diamond with a particle size ranging between 53 and 63 μm. A Ti6Al4V alloy powder was used as the matrix material. The diamond reinforcement weight percentage was varied from 10 to 50 wt.%. The layer thickness was kept at 0.254 mm, whereas the scanning speed was varied according to the different composite specimens. The compressive strength, hardness, and elastic modulus of the composite materials were investigated. The greatest hardness value was discovered for the titanium alloy composite specimen with 40 wt.% diamond particles, after which there was a noticeable reduction in the hardness value for 50 wt.% diamond reinforcement. Furthermore, the highest compressive strength value of 1669 MPa was noted for 20 wt.% diamond composites, whereas the maximum tensile modulus value of 131 GPa was found for 30 wt.% diamond composites. After that, both the tensile modulus and compressive

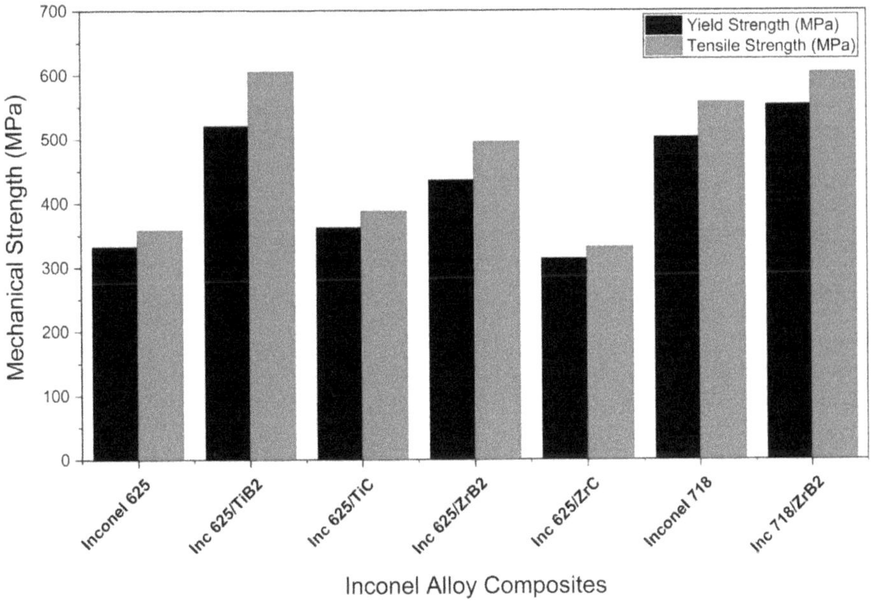

FIGURE 10.7 Mechanical properties of Inconel 625 and 718 alloys along with their composites.

strength were decreased with additional diamond loading (Chen et al., 2018). Another study explored the utilization of graphene as reinforcement in fabricating MMCs. Firstly, a mixture of graphene and aluminum powder was created by the ball milling method with different graphene content weight percentages of 0.5, 1.0, and 2.5%. After that, SLM was used to develop these composites. The mechanical behavior of these composites was analyzed and compared with pure Al. The hardness was evaluated using the Vickers hardness test, whereas nanomechanical properties were assessed using Nano Intender G200. The greatest hardness value was noted for composites with 2.5 wt.% graphene content. This was 75.3% better than pure Al. Furthermore, the nanomechanical properties were determined, and the nano hardness value was found to be 1.02 and 1.75 GPa for pure Al and composites, respectively, with 2.5 wt.% graphene. The research showed the utilization of graphene as a promising reinforcement in fabricating 3D MMCs with enhanced mechanical properties (Hu et al., 2018). Figure 10.8 shows the hardness value of titanium and aluminum alloy MMCs. A summary of the mechanical behavior of MMCs is reflected in Table 10.3.

10.3.2 Thermal Behavior of MMCs

MMCs exhibit good thermal stability in comparison to PMCs. MMCs display high thermal conductivity, especially with aluminum as the matrix. Residual thermal stress may develop due to uneven or rapid cooling in laser-based AM processes. Additionally, there is a risk of porosity and microcracks forming due to poor process

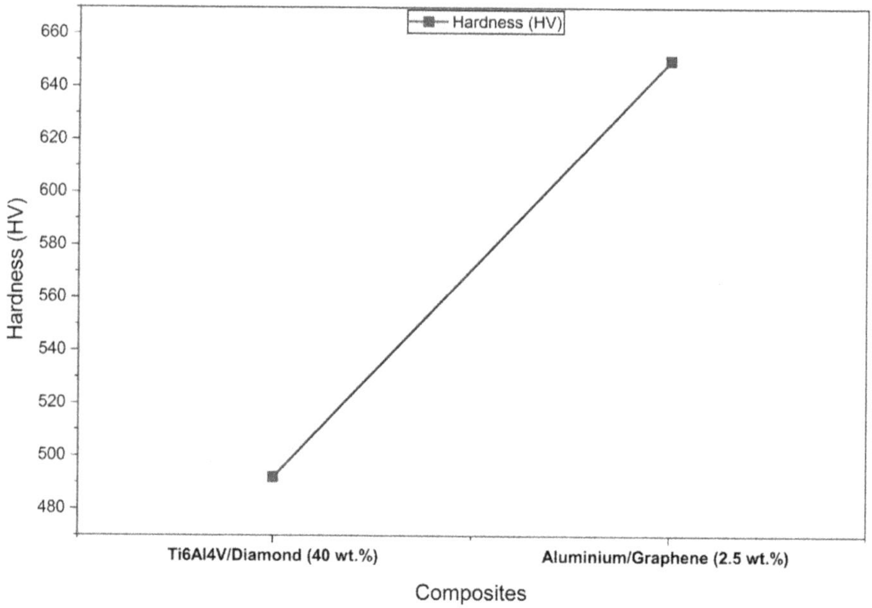

FIGURE 10.8 Hardness value of titanium and aluminum alloy metal matrix composites (Chen et al., 2023; Hu et al., 2018).

control. Generally, 3D-printed MMCs require post-processing treatment like HIP or annealing to relieve induced thermal stress and improve overall thermal performance (Kumar et al., 2022). The authors have observed that there are only a few literature studies available that discuss the thermal behavior of 3D-printed MMCs in comparison to their mechanical behavior. Thus, the few studies that are available have been discussed further.

A study analyzed the thermal behavior of aluminum alloy/SiC MMCs, which were created by LPBF. The weight percentages of SiC reinforcement used was 2 and 5 wt.%, respectively. Thermal behavior, like the thermal expansion and thermal conductivity of the composite materials, were studied. The thermal conductivity of 3D-printed specimens was analyzed using laser flash diffusivity. The temperature range for thermal conductivity was between 25 and 150°C, with an interval of 25°C. It was observed that the incorporation of SiC particles in an aluminum alloy at both 2 and 5 wt.% lowered the composites' thermal conductivity in comparison to a neat aluminum alloy. The reason behind this is the low thermal conductivity of the SiC particles themselves. The linear thermal expansion of the composite specimens was analyzed utilizing a dual push-rod dilatometer. It was revealed that Al/SiC composites with 2 wt.% SiC had a high coefficient of thermal expansion at each temperature interval in comparison to the composites with 5 wt.% SiC, although the change was not very significant (Bai & Liu, 2024). Titanium alloys are known for their exceptional mechanical strength, but their low thermal conductivity hinders their broader applications. Thus, a study utilized DED to fabricate titanium (Ti6Al4V)/diamond

TABLE 10.3
Mechanical Properties of MMCs

Material	Matrix/ fiber content	Method	Printing speed (mm/s)	Layer thickness (mm)	Tensile strength (MPa)	Yield strength (MPa)	Breaking elongation (%)	Reference
Aluminum alloy (2024-T3)/ carbon fiber-reinforced onyx composite	50 % volume fraction of composites	FDM	-	0.125	240.0	-	-	Yelamanchi et al., 2020
Bronze/ alumina	-	Binder jetting and TDS	30.0	0.1-0.5	154.0	110.0	-	Yoozbashizadeh et al., 2018
Aluminum alloy (AlSi10Mg)/ SiC	2 wt.% SiC	LPBF	100.0	0.05	403.0	221.0	5.0	Bai & Liu, 2024
Aluminum alloy (AA2024)/ aluminum oxide	25 wt.% aluminum oxide	Cold spray 3D printing	-	-	290.0	-	1.4	Wu et al., 2022
Inconel 625/TiB_2	2 wt.% TiB_2	Laser powder bed fusion	-	-	605.0	520.0	7.2	Tekoğlu et al., 2024
Inconel 718/ZrB_2	2 wt.% ZrB_2	Laser powder bed fusion	-	0.04	603.1	552.2	9.2	Tekoğlu et al., 2024
Aluminum alloy/ alumina fibers	35 wt.% alumina fibers	FFF	-	-	269.0	-	-	Seleznev et al., 2023
PbSn alloy/ Cu-coated carbon fibers	50 wt.% Pb and Sn	Novel 3D printing method	30.0	0.3	236.7	-	-	Wang et al., 2020

(continued)

TABLE 10.3 (Continued)
Mechanical Properties of MMCs

Material	Matrix/ fiber content	Method	Printing speed (mm/s)	Layer thickness (mm)	Tensile strength (MPa)	Yield strength (MPa)	Breaking elongation (%)	Reference
Cu174PH	35 wt.% Cu	LPBF	200.0	0.03	608.0	-	-	Lusicini et al., 2025
Continuous carbon fiber/bronze	-	MEX	50	0.4	184.0	65.0	-	Mousapour et al., 2025
TiC/nickel alloy (In738LC)	2 wt.% TiC	LPBF	-	-	1416.0	1020.0	3.26	Shu et al., 2025

MMCs with improved thermal conductivity. The diamond weight percentage was varied from 10 to 50 wt.%. The thermal behavior of the composites was enhanced with the inclusion of diamond particles. Both specific heat and thermal conductivity were observed to increase with increasing diamond loading in comparison to pure titanium alloy. The thermal conductivity was improved by 200% for composites with 40 wt.% diamond loading. At 50 wt.% diamond loading, the thermal conductivity was found to decrease, but it was still better than the pure titanium alloy (Chen et al., 2023).

Another research study explored the thermal behavior of aluminum alloy/carbon fiber metal matrix composites, which were developed by SLM using a mixture of aluminum and short carbon fiber powder. To enhance the thermal behavior of the composite materials, carbon fiber-coated Cu powder was mixed with Al powder to create Cu-coated carbon fiber/Al composites. The composites were created to achieve anisotropic thermal conductivity (ATC). The composite samples were manufactured with a unidirectional type of scanning pattern. The thermal conductivity of the 0° direction composite sample was noted to be twice as low as the 90° direction composite sample. Moreover, ATC was found in these composites in comparison to Cu-coated CF/Al composites prepared with a rot pattern (Kimura et al., 2020). A study utilized SLM to develop pure aluminum alloy (AlSi10Mg) specimens to investigate their thermal conductivity. The influence of different processing parameters, like build direction, hatch distance, and laser scanning velocity on the thermal conductivity of the samples was analyzed. The experimental outcomes of different overhang angles showed that the thermal conductivity of the composite sample with a 90° overhang angle was higher than the sample with a 0° overhang angle. Considering the laser scan speed, it was summarized that, as the scan speed was increased, the specimens displayed a reduction in thermal conductivity. Hatch distances of 0.1, 0.12, and 0.14 mm were used. It was expressed that the thermal conductivity was higher when the hatch distance was shorter (Kim , 2021). The thermal conductivity of MMCs is summarized in Figure 10.9.

Another research study examined the thermal behavior of novel Cu174PH AM metal matrix composites. This innovative material was obtained by mixing 65% pure copper and 35% 17-4 PH (precipitation-hardened stainless steel). The composites were fabricated using LPBF. The application of various heat treatment methods on the thermal behavior of the composite specimens was examined. The heat treatment process was carried out in a proportional integral derivative (PID)-controlled oven. The heat treatment was done at various temperatures, such as 400, 500, 600, and 700°C for a period of ten hours. Finally, the quenching process was used to test its influence on the MMCs' properties. The age-hardening heat treatment of 500, 600, and 700°C for ten hours provided a good balance of thermal conductivity. The highest thermal conductivity was observed for composite samples that had undergone heat treatment at a temperature of 600°C for ten hours (Lusicini et al., 2025). Analyzing the thermal behavior of MMCs is an important aspect for various applications in the automotive, aerospace, and space industries. Future work should focus more on the thermal characterization of AM MMCs in correlation with mechanical behavior.

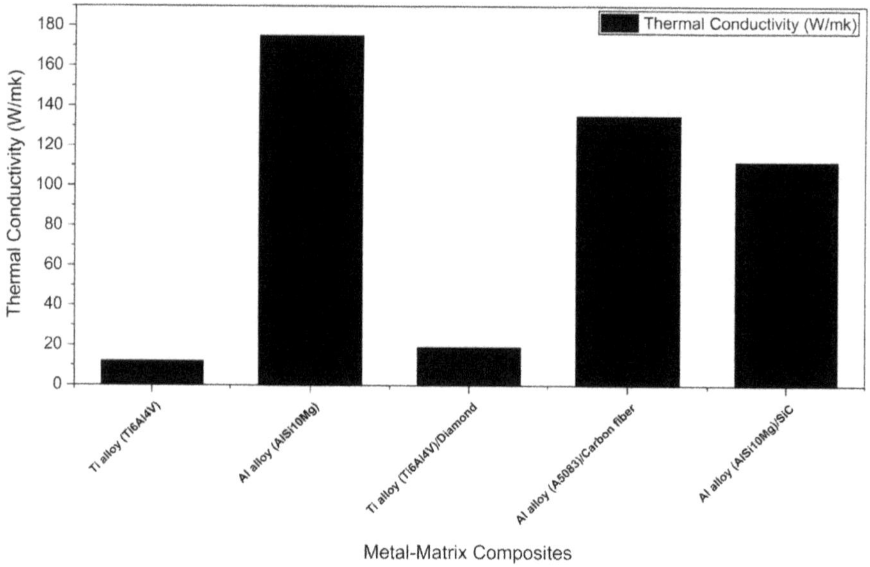

FIGURE 10.9 Thermal conductivity of MMCs in the temperature range of 100°C to 150°C.

10.4 THERMO-MECHANICAL BEHAVIOR OF CERAMIC MATRIX COMPOSITES

10.4.1 Mechanical Behavior of CMCs

CMCs are known for their excellent stability at high temperatures, high resistance to wear, high strength, good fracture toughness, and good thermal shock resistance. The utilization of AM to fabricate ceramics and CMCs is still in an early phase in comparison to metals and polymers. Commonly used 3D-printing techniques to produce ceramics include SLA, SLS, digital light processing (DLP), binder jetting, and material extrusion. These methods can create complex shapes, and the final properties of materials can be similar to those obtained from traditional ceramic manufacturing methods. Only a few studies have been conducted on 3D-printed CMCs as materials are hard to print and sinter. The 3D printing of CMCs requires multiple complex steps for both fibers and matrix. However, a few studies have developed novel 3D-printing methods or used a combination of methods to create CMCs. These are discussed further.

A study investigated the mechanical behavior of CMCs, which were fabricated by SLS, followed by polymer infiltration and pyrolysis (PIP). Here, the matrix material used was SiC powder, while the reinforcement utilized was short carbon fibers, leading to the fabrication of dense SiC/carbon fiber composites. The percentage of carbon fibers was varied from 0 to 12 wt.%, whereas a diameter of 70 microns and lengths of 30, 50, and 70 micron-sized fibers were used. E12 epoxy powder was utilized as a binding agent to bond the SiC and carbon fiber. The printing powder consisted of SiC, carbon fiber, and E12 epoxy powder. The volume fraction of epoxy powder was kept at 5%. The impact of different carbon fiber sizes and loading on the properties of the composite samples was examined. The fracture toughness of composites increased, but then decreased as the fiber loading increased. It was found that SiC/carbon fiber composites with 8 wt.% carbon fibers and a 70 μm length had a 33.3% increase in fracture toughness compared to a neat SiC matrix. The highest bending strength was observed for composites with fibers at 4 wt.% and a 70 μm length at 138.9 MPa, although it was less than the neat SiC matrix composite, which had a bending strength of 140.8 MPa (Zhang et al., 2024). In other research, 3D-printed CMCs with hierarchical structures were fabricated. Short SiC fibers were successfully added to silicate materials through the formulation of printing precursors. The 3D-printed parts exhibited a gradient porous hierarchical structure. Three different types of filler materials (SiC, Al_2O_3, and ZrO_2) were used as reinforcement. The composite structure was additively manufactured utilizing aluminum silicate powder as the matrix material. The matrix powder was mixed with filler material in the ratio of 50:1. After mixing was completed, water was included in the powder to form a paste (water-to-powder ratio of 6:10) for utilization in 3D printing. For composites with SiC as filler material, the impact strength, bending strength, and bending modulus were enhanced by 27, 7.5, and 22%. For composites with other filler materials, such as Al_2O_3 and ZrO_2, there was no noticeable enhancement in the mechanical performance (Chua et al., 2022).

Another research study utilized a combination of binder jetting, spray drying, and chemical vapor infiltration (CVI) methods to manufacture SiC whisker-reinforced SiC CMCs. By using the spray drying technique, the SiC whiskers, having poor flowability and an irregular shape, were turned into a spherical shape due to their spherical particles being a favorable material for the binder jetting process. SiC whisker preforms with two-stage pores were created by binder jetting. Additionally, the SiC matrix created by CVI was dense and pure, with great strength and rigidity. The incorporation of an SiC matrix enabled the consolidation of the SiC matrix. The crack toughness (fracture toughness) and bending strength of the fabricated composites were evaluated. The bending strength was noted to be 200 MPa, whereas fracture toughness was found to be 3.4 MPa $m^{1/2}$ (Lv et al., 2019). Another study created silicon oxycarbide (SiOC) ceramic matrix composites utilizing DLP by photopolymerizing a siloxane-based pre-ceramic polymer matrix and including inert reinforcement. The heat treatment process was employed to convert pre-ceramic polymer parts into an SiOC ceramic matrix. The SiC whiskers, and particles such as Al_2O_3, Si_3N_4, and mullite were used as reinforcements to create ceramic composites. The inclusion of reinforcement had a positive effect on the composite's properties. The toughness of the composites for both the particles and whisker reinforcement was increased to above 3 MPa $m^{1/2}$ in contrast to a pristine SiOC matrix, which had a toughness value of 1 MPa $m^{1/2}$. The bending strength of the ceramic composites was also found to increase from 225 to 335 MPa. This research showed that the 3D-printing technique that was developed allowed for the free-form development of complex and high-performance composite samples (O'Masta et al., 2020). A study utilized material extrusion-based 3D printing to manufacture oxide ceramic matrix composites. The reinforcement used was alumina fibers, which was introduced in a

TABLE 10.4
Flexural Strength of a Few CMCs

Matrix	Reinforcement	Reinforcement content	Flexural strength (MPa)	Reference
Silicon carbide (SiC)	Carbon fiber	4 wt.%	135.0	Zhang et al., 2024
Aluminum silicate	Short SiC fibers	2 wt.%	21.74	Chua et al., 2022
Silicon carbide (SiC)	SiC whiskers	17.7 wt.%	200.0	Lv et al., 2019
Silicon oxycarbide (SiOC)	Mullite	15 wt.%	270.0	O'Masta et al., 2020
Zirconia-toughened alumina powder	Alumina	10 wt.%	108.0	Hadian et al., 2024

powdered mixture of zirconia-toughened alumina, and the filaments were extruded. The fiber volume was varied from 0 to 30 wt.%. The mechanical behavior of filaments and additively manufactured specimens was examined. The extruded filaments with 15 wt.% alumina fibers had the highest bending strength of 110 MPa. A further rise in fiber loading led to a decline in the bending strength of the filaments. Using Weibull analysis, it was discovered that the 3D-printed specimens without fibers had a lower flexural strength in comparison to the extruded filaments owing to printing factors such as voids and stresses. It was also demonstrated that printed composite specimens with 10 wt.% alumina fibers and a 0° orientation achieved the same flexural strength as the extruded filaments. The results showed that using fibers as reinforcement could be a feasible solution to avoid a reduction in mechanical properties due to printing defects (Hadian et al., 2024). Table 10.4 provides a summary of the flexural strength of CMCs.

10.4.2 Thermal Behavior of CMCs

CMCs are known for having high temperature resistance, typically above 1000°C. CMCs have a very low coefficient of thermal expansion, owing to their superior dimensional stability under thermal cyclic loading. By using carbon or SiC particles as reinforcement, the thermal conductivity of CMCs can be increased. In 3D printing, porosity due to incomplete sintering may reduce the thermal performance of composites. Therefore, high-temperature post-processing of CMCs is required for densification (Derby, 2015; Peng et al., 2018). The literature on the development and characterization of AM CMCs is scarce, let alone on the thermal behavior of these composites. However, some combination of printing and post-processing methods alters the microstructure and thus influences the thermal behavior of CMCs. This is addressed further.

Additively manufactured SiC-based ceramic matrix composites display similar benefits to traditional CMCs, where these composites can be used at elevated temperatures (>1000°C), but the desired service temperature and long-term reliability strongly depend on the printing technique used, printing process parameters, and post-processing methods, such as CVI, sintering, and densification. The thermal conductivity of 3D-printed CMCs is strongly dependent on factors like fiber and matrix interaction, the fiber orientation/architecture introduced by the 3D printing process, and the residual porosity and secondary phases from printing/post-processing (Byun et al., 2021; Sun et al., 2023). A few reported studies of the thermal conductivity of 3D-printed SiC depict spatial anisotropy along the fiber direction. Additively manufactured CMCs display good thermal shock resistance if printing and post-processing parameters preserve toughening mechanisms (crack deflection, fiber bridging). However, when heated or cooled rapidly, porous or weakly bonded printed microstructures can increase spallation and decrease thermal shock resistance. Recent studies have explicitly tested for thermal shock in certain porous 3D-printed ceramics, such as zirconia lattices (Costa Oliveira et al., 2023; Pek et al., 2020).

A chosen AM route and post-processing method strongly controls the microstructure of CMCs, thereby influencing their thermal behavior. A study showed that combining binder jet 3D printing and silicon melt infiltration can yield near-SiC composites, but microstructural heterogeneity affected thermal conductivity and oxidation (Cramer et al., 2021). Another research study combined automated fiber placement, pyrolysis, and CVI techniques to produce continuous carbon fiber preforms, with CMCs having a microstructure very similar to traditional CMCs, and thus displaying better thermal/mechanical performance (Cramer et al., 2022). It can be concluded that thermal behavior is not intrinsic to "3D-printed CMCs" generically, but is mostly process dependent. Only a small number of published research studies have experimentally investigated the long-term creep or high-temperature life of 3D-printed CMCs. For printed SiC, oxidation and short-term thermophysical testing are more prevalent than long-duration creep investigations; the majority of creep/thermomechanical research still focuses on conventionally made CMCs. There is a significant gap in the literature here, which must be explored in the future.

10.5 CONCLUSION

This chapter has provided insights into the mechanical and thermal behavior of PMCs, MMCs, and CMCs, which are fabricated using 3D printing methods like FDM, SLA, SLS, and others. Additive manufacturing has emerged as an advanced manufacturing technique in recent years, capable of creating functional end-user products. 3D-printing techniques, known for their innovation and versatility, provide significant benefits across different industrial sectors, such as the automotive, aviation, electronics, healthcare, civil engineering, and fashion industries. The 3D-printing technique is used for different applications, ranging from making simple electronic and plastic components to more complex real-world applications such as human organs, automotive structural components, and large-scale houses. With respect to research gaps, the authors could not find any article that demonstrates experimental characterization, such as tensile properties and the creep resistance of 3D-printed CMCs, which could serve as a potential for future research. This review will help researchers, who are advancing in the field of AM, by providing information on the properties of 3D-printed materials. In summary, AM is on the rise and is the future of manufacturing, with the ability to change how parts are designed, created, and supplied.

REFERENCES

Andreozzi, M., Gentili, S., Forcellese, P., Bellezze, T., Corinaldesi, V., Luzi, F., & Vita, A. (2024). Effect of moisture content on the mechanical performance of 3D printed continuous reinforced two-matrix composite. *The International Journal of Advanced Manufacturing Technology*, *133*(11), 5117–5126. https://doi.org/10.1007/s00170-024-14041-5

Aramian, A., Razavi, N., Sadeghian, Z., & Berto, F. (2020). A review of additive manufacturing of cermets. *Additive Manufacturing*, *33*, 101130. https://doi.org/10.1016/j.addma.2020.101130

Arora, N., Dua, S., Choudhary, V., Singh, S. K., & Senthilkumar, T. (2025). Mechanical properties of novel PLA composite infused with betel nut waste biocarbon for sustainable 3D printing. *Composites Communications*, *53*, 102188. https://doi.org/10.1016/j.coco.2024.102188

Bai, S., & Liu, J. (2024). Empirical investigation of properties for additive manufactured aluminum metal matrix composites. *Applied Mechanics*, *5*(3), 450–474. https://doi.org/10.3390/applmech5030026

Bute, I., Tarasovs, S., Vidinejevs, S., Vevere, L., Sevcenko, J., & Aniskevich, A. (2023). Thermal properties of 3D printed products from the most common polymers. *The International Journal of Advanced Manufacturing Technology*, *124*(7), 2739–2753. https://doi.org/10.1007/s00170-022-10657-7

Byun, T. S., Lach, T. G., Parish, C. M., Wang, H., Trofimov, A. A., Collins, D. A., ... & Terrani, K. A. (2021). *Mechanical and Thermophysical Properties of 3D-Printed SiC Before and After Neutron Irradiation–FY21* (No. ORNL/TM-2021/2006). Oak Ridge National Laboratory (ORNL), Oak Ridge, TN (United States). https://doi.org/10.2172/1814305

Chen, L., Wilson, R., de Looze, G., Yang, K., & Seeber, A. (2023). Additive manufacturing of titanium alloy based composites using directed energy deposition: Study of microstructure, mechanical properties and thermal conductivity. *Materials Science and Engineering: A*, *884*, 145579. https://doi.org/10.1016/j.msea.2023.145579

Chua, C. S., Ow, S. Y., Liew, S. L., Liu, J., Soh, C. B., Shen, L., & Wang, S. (2022). 3D printing of fibre-reinforced ceramic composites with hierarchical structure. *Advances in Applied Ceramics*, *121*(2), 46–51. https://doi.org/10.1080/17436753.2021.2014276

Coelho, A. W. F., Thiré, R. M. D. S. M., & Araujo, A. C. (2019). Manufacturing of gypsum–sisal fiber composites using binder jetting. *Additive Manufacturing*, *29*, 100789. https://doi.org/10.1016/j.addma.2019.100789

Costa Oliveira, F. A., Sardinha, M., Galindo, J., Rodríguez, J., Cañadas, I., Leite, M., & Fernandes, J. C. (2023). Manufacturing and thermal shock resistance of 3D-printed porous black zirconia for concentrated solar applications. *Crystals*, *13*(9), 1323. https://doi.org/10.3390/cryst13091323

Cramer, C. L., Elliott, A. M., Lara-Curzio, E., Flores-Betancourt, A., Lance, M. J., Han, L., ... & Nawaz, K. (2021). Properties of SiC-Si made via binder jet 3D printing of SiC powder, carbon addition, and silicon melt infiltration. *Journal of the American Ceramic Society, 104*(11), 5467–5478. https://doi.org/10.1111/jace.17933

Cramer, C. L., Yoon, B., Lance, M. J., Cakmak, E., Campbell, Q. A., & Mitchell, D. J. (2022). Additive manufacturing of C/C-SiC ceramic matrix composites by automated fiber placement of continuous fiber tow in polymer with pyrolysis and reactive silicon melt infiltration. *Journal of Composites Science, 6*(12), 359. https://doi.org/10.3390/jcs6120359

Dadkhah, M., Mosallanejad, M. H., Iuliano, L., & Saboori, A. (2021). A comprehensive overview on the latest progress in the additive manufacturing of metal matrix composites: Potential, challenges, and feasible solutions. *Acta Metallurgica Sinica (English Letters), 34*(9), 1173–1200. https://doi.org/10.1007/s40195-021-01249-7

Derby, B. (2015). Additive manufacture of ceramics components by inkjet printing. *Engineering, 1*(1), 113–123. https://doi.org/10.15302/J-ENG-2015014

Fu, X., Lin, Y., Yue, X. J., XunMa, Hur, B., & Yue, X. Z. (2022, May). A review of additive manufacturing (3D printing) in aerospace: Technology, materials, applications, and challenges. In *Mobile Wireless Middleware, Operating Systems and Applications: 10th International Conference on Mobile Wireless Middleware, Operating Systems and Applications (MOBILWARE 2021)* (pp. 73–98). Springer International Publishing. https://doi.org/10.1007/978-3-030-98671-1_6

Gauss, C., Pickering, K. L., Graupner, N., & Müssig, J. (2023). 3D-printed polylactide composites reinforced with short lyocell fibres–Enhanced mechanical properties based on bio-inspired fibre fibrillation and post-print annealing. *Additive Manufacturing, 77*, 103806. https://doi.org/10.1016/j.addma.2023.103806

Gupta, A., Fidan, I., Hasanov, S., & Nasirov, A. (2020). Processing, mechanical characterization, and micrography of 3D-printed short carbon fiber reinforced polycarbonate polymer matrix composite material. *The International Journal of Advanced Manufacturing Technology, 107*(7), 3185–3205. https://doi.org/10.1007/s00170-020-05195-z

Hadian, A., Duckek, J., Parrilli, A., Liersch, A., & Clemens, F. (2024). Additive manufacturing of fiber-reinforced zirconia-toughened alumina ceramic matrix composites by material extrusion-based technology. *Advanced Engineering Materials, 26*(18), 2302158. https://doi.org/10.1002/adem.202302158

Hu, Z., Chen, F., Xu, J., Nian, Q., Lin, D., Chen, C., ... & Zhang, M. (2018). 3D printing graphene-aluminum nanocomposites. *Journal of Alloys and Compounds, 746*, 269–276. https://doi.org/10.1016/j.jallcom.2018.02.272

Hwang, S., Fowler, G., Han, Y., & Gardner, D. J. (2024). Material property characterization of 3D printed polypropylene wood plastic composites. *Polymer Composites, 45*(17), 16058–16074. https://doi.org/10.1002/pc.28890

Hwang, S., Reyes, E. I., Moon, K. S., Rumpf, R. C., & Kim, N. S. (2015). Thermo-mechanical characterization of metal/polymer composite filaments and printing parameter study for fused deposition modeling in the 3D printing process. *Journal of Electronic Materials, 44*(3), 771–777. https://doi.org/10.1007/s11664-014-3425-6

Ingrassia, T., Nigrelli, V., Ricotta, V., & Tartamella, C. (2016, September). Process parameters influence in additive manufacturing. In *Advances on Mechanics, Design Engineering and Manufacturing: Proceedings of the International Joint Conference on Mechanics, Design Engineering & Advanced Manufacturing (JCM 2016), 14-16 September, 2016, Catania, Italy* (pp. 261–270). Springer International Publishing. https://doi.org/10.1007/978-3-319-45781-9_27

Jadhav, A., & Jadhav, V. S. (2022). A review on 3D printing: An additive manufacturing technology. *Materials Today: Proceedings*, *62*, 2094–2099. https://doi.org/10.1016/j.matpr.2022.02.558

Jaganathan, S., Kandasamy, R., Venkatachalam, R., Gunalan, M., & Dhairiyasamy, R. (2024). Advances in optimizing mechanical performance of 3D-printed polymer composites: A microstructural and processing enhancements review. *Advances in Polymer Technology*, *2024*(1), 3168252. https://doi.org/10.1155/2024/3168252

Jung, J. Y., Yu, S., Kim, H., Cha, E., Shin, G. S., Eo, S. B., ... & Hwang, J. Y. (2025). Process-structure–property study of 3D-printed continuous fiber reinforced composites. *Composites Part A: Applied Science and Manufacturing*, *188*, 108538. https://doi.org/10.1016/j.compositesa.2024.108538

Kilinc, F. B., Bozaci, E., Kilinc, A. C., & Turkoglu, T. (2025). Effect of atmospheric plasma treatment on mechanical properties of 3D-printed continuous aramid fiber/PLA composites. *Polymers*, *17*(3), 397. https://doi.org/10.3390/polym17030397

Kim, M. S. (2021). Effects of processing parameters of selective laser melting process on thermal conductivity of alsi10mg alloy. *Materials*, *14*(9), 2410. https://doi.org/10.3390/ma14092410

Kimura, T., Nakamoto, T., Suyama, T., & Miki, T. (2020). In-process fabrication of carbon-dispersed aluminum matrix composite using selective laser melting. *Metals*, *10*(5), 619. https://doi.org/10.3390/met10050619

Kumar, R., Kumar, S., Kumar, M., & Luthra, G. (2022). Conventional and 3D printing technology for the manufacturing of metal-matrix composite: A study. In Bansal, S. A., Khanna, V., & Gupta, P. (Eds.), *Metal Matrix Composites* (pp. 197–223). CRC Press. https://doi.org/10.1201/9781003194897

Laureto, J., Tomasi, J., King, J. A., & Pearce, J. M. (2017). Thermal properties of 3-D printed polylactic acid-metal composites. *Progress in Additive Manufacturing*, *2*(1), 57–71. https://doi.org/10.1007/s40964-017-0019-x

Liu, X., Tey, W. S., Choo, J. Y. C., Chen, J., Tan, P., Cai, C., ... & Zhou, K. (2021). Enhancing the mechanical strength of multi jet fusion–printed polyamide 12 and its glass fiber-reinforced composite via high-temperature annealing. *Additive Manufacturing*, *46*, 102205. https://doi.org/10.1016/j.addma.2021.102205

Lusicini, D., Crachi, M., Sesana, R., Delprete, C., & Sicignano, N. (2025). Experimental analysis of thermomechanical properties of innovative heat treated metal matrix composites processed by additive manufacturing for space rocket applications. *Journal of Materials Engineering and Performance*, 34, 19824–19839. https://doi.org/10.1007/s11665-025-11165-x

Lv, X., Ye, F., Cheng, L., Fan, S., & Liu, Y. (2019). Fabrication of SiC whisker-reinforced SiC ceramic matrix composites based on 3D printing and chemical vapor infiltration technology. *Journal of the European Ceramic Society*, *39*(11), 3380–3386. https://doi.org/10.1016/j.jeurceramsoc.2019.04.043

Mostafaei, A., Elliott, A. M., Barnes, J. E., Li, F., Tan, W., Cramer, C. L., ... & Chmielus, M. (2021). Binder jet 3D printing—Process parameters, materials, properties, modeling, and challenges. *Progress in Materials Science*, *119*, 100707. https://doi.org/10.1016/j.pmatsci.2020.100707

Mousapour, M., Kumar, S. S., Partanen, J., & Salmi, M. (2025). 3d printing of a continuous carbon fiber reinforced bronze-matrix composite using material extrusion. *Composites Part B: Engineering*, *289*, 111961. https://doi.org/10.1016/j.compositesb.2024.111961

Mubarak, S., Dhamodharan, D., B. Kale, M., Divakaran, N., Senthil, T., Sathiyanathan, P.,... & Wang, J. (2020). A novel approach to enhance mechanical and thermal properties of SLA 3D printed structure by incorporation of metal–metal oxide nanoparticles. *Nanomaterials*, *10*(2), 217. https://doi.org/10.3390/nano10020217

Murr, L. E. (2018). A metallographic review of 3D printing/additive manufacturing of metal and alloy products and components. *Metallography, Microstructure, and Analysis*, *7*(2), 103–132. https://doi.org/10.1007/s13632-018-0433-6

Mushtaq, R. T., Iqbal, A., Wang, Y., Rehman, M., & Petra, M. I. (2023). Investigation and optimization of effects of 3D printer process parameters on performance parameters. *Materials*, *16*(9), 3392. https://doi.org/10.3390/ma16093392

Nurazzi, N., Asyraf, M. R. M., Rayung, M., Norrrahim, M. N. F., Shazleen, S. S., Rani, M. S. A., ... & Abdan, K. (2021). Thermogravimetric analysis properties of cellulosic natural fiber polymer composites: A review on influence of chemical treatments. *Polymers*, *13*(16), 2710. https://doi.org/10.3390/polym13162710

Ogaili, A. A. F., Basem, A., Kadhim, M. S., Al-Sharify, Z. T., Jaber, A. A., Njim, E. K., ... & Al-Ameen, E. S. (2024). The effect of chopped carbon fibers on the mechanical properties and fracture toughness of 3D-printed PLA parts: An experimental and simulation study. *Journal of Composites Science*, *8*(7), 273. https://doi.org/10.3390/jcs8070273

O'Masta, M. R., Stonkevitch, E., Porter, K. A., Bui, P. P., Eckel, Z. C., & Schaedler, T. A. (2020). Additive manufacturing of polymer-derived ceramic matrix composites. *Journal of the American Ceramic Society*, *103*(12), 6712–6723. https://doi.org/10.1111/jace.17275

Parandoush, P., & Lin, D. (2017). A review on additive manufacturing of polymer-fiber composites. *Composite Structures*, *182*, 36–53. https://doi.org/10.1016/j.compstruct.2017.08.088

Parupelli, S., & Desai, S. (2019). A comprehensive review of additive manufacturing (3D printing): Processes, applications and future potential. *American Journal of Applied Sciences*, *16*(8), 244–272. https://thescipub.com/abstract/ajassp.2019.244.272.

Pek, E. K., Brethauer, J., & Cahill, D. G. (2020). High spatial resolution thermal conductivity mapping of SiC/SiC composites. *Journal of Nuclear Materials*, *542*, 152519. https://doi.org/10.1016/j.jnucmat.2020.152519

Peng, E., Zhang, D., & Ding, J. (2018). Ceramic robocasting: Recent achievements, potential, and future developments. *Advanced Materials*, *30*(47), 1802404. https://doi.org/10.1002/adma.201802404

Prabhakar, M. M., Saravanan, A. K., Lenin, A. H., Mayandi, K., & Ramalingam, P. S. (2021). A short review on 3D printing methods, process parameters and materials. *Materials Today: Proceedings*, *45*, 6108–6114. https://doi.org/10.1016/j.matpr.2020.10.225

Quill, T. J., Smith, M. K., Zhou, T., Baioumy, M. G. S., Berenguer, J. P., Cola, B. A., ... & Bougher, T. L. (2018). Thermal and mechanical properties of 3D printed boron nitride–ABS composites. *Applied Composite Materials*, *25*(5), 1205–1217. https://doi.org/10.1007/s10443-017-9661-1

Saba, N., Jawaid, M., Alothman, O. Y., & Paridah, M. T. (2016). A review on dynamic mechanical properties of natural fibre reinforced polymer composites. *Construction and Building Materials*, *106*, 149–159. https://doi.org/10.1016/j.conbuildmat.2015.12.075

Seleznev, M. L., Roy-Mayhew, J. D., & Faust, J. L. (2023). Fabrication and tunable reinforcement of net-shaped aluminum matrix composite parts via 3D printing. *Scientific Reports*, *13*(1), 16334. https://doi.org/10.1038/s41598-023-43514-y

Shi, K., Yan, Y., Mei, H., Chen, C., & Cheng, L. (2021). 3D printing Kevlar fiber layer distributions and fiber orientations into nylon composites to achieve designable mechanical strength. *Additive Manufacturing*, *39*, 101882. https://doi.org/10.1016/j.addma.2021.101882

Shu, C., Chen, S., Lei, P., Shu, X., Abdel-Wahab, A., & Essa, K. (2025). Effect and mechanism of powder morphology on mechanical properties of nickel-based metal matrix composites TiC-IN738LC in laser powder bed fusion. *Progress in Additive Manufacturing*, 10, 7835–7853. https://doi.org/10.1007/s40964-025-01075-8

Sun, J., Ye, D., Zou, J., Chen, X., Wang, Y., Yuan, J., ... & Bai, J. (2023). A review on additive manufacturing of ceramic matrix composites. *Journal of Materials Science & Technology*, *138*, 1–16. https://doi.org/10.1016/j.jmst.2022.06.039

Tekinalp, H. L., Kunc, V., Velez-Garcia, G. M., Duty, C. E., Love, L. J., Naskar, A. K., ... & Ozcan, S. (2014). Highly oriented carbon fiber–polymer composites via additive manufacturing. *Composites Science and Technology*, *105*, 144–150. https://doi.org/10.1016/j.compscitech.2014.10.009

Tekoglu, E., Bae, J. S., Kim, H. A., Lim, K. H., Liu, J., Doležal, T. D., ... & Li, J. (2024). Superior high-temperature mechanical properties and microstructural features of LPBF-printed In625-based metal matrix composites. *Materials Today*, *80*, 297–307. https://doi.org/10.1016/j.mattod.2024.09.006

Tekoğlu, E., O'Brien, A. D., Bae, J. S., Lim, K. H., Liu, J., Kavak, S., ... & Li, J. (2024). Metal matrix composite with superior ductility at 800° C: 3D printed In718+ ZrB2 by laser powder bed fusion. *Composites Part B: Engineering*, *268*, 111052. https://doi.org/10.1016/j.compositesb.2023.111052

Tikhani, F., Gurbin, A., & Hubert, P. (2024). Unveiling the impact of short fibre reinforcement and extrusion properties on microstructure of 3D printed polycarbonate composites. *Additive Manufacturing*, *93*, 104423. https://doi.org/10.1016/j.addma.2024.104423

Torrado Perez, A. R., Roberson, D. A., & Wicker, R. B. (2014). Fracture surface analysis of 3D-printed tensile specimens of novel ABS-based materials. *Journal of Failure Analysis and Prevention*, *14*(3), 343–353. https://doi.org/10.1007/s11668-014-9803-9

Torrado, A. R., Shemelya, C. M., English, J. D., Lin, Y., Wicker, R. B., & Roberson, D. A. (2015). Characterizing the effect of additives to ABS on the mechanical property anisotropy of specimens fabricated by material extrusion 3D printing. *Additive Manufacturing*, *6*, 16–29. https://doi.org/10.1016/j.addma.2015.02.001

Wang, X., Tian, X., Yin, L., & Li, D. (2020). 3D Printing of continuous fiber reinforced low melting point alloy matrix composites: Mechanical properties and microstructures. *Materials*, *13*(16), 3463. https://doi.org/10.3390/ma13163463

Wang, Y., Li, F., Sun, Z., Gu, C., Fu, K., & Zhao, X. (2025). Effect of fiber content on the preparation and mechanical properties of 3D printed short carbon fiber reinforced PA composites. *Polymers*, *17*(5), 671. https://doi.org/10.3390/polym17050671

Wu, D., Li, W., Liu, K., Yang, Y., & Hao, S. (2022). Optimization of cold spray additive manufactured AA2024/Al2O3 metal matrix composite with heat treatment. *Journal of Materials Science & Technology*, *106*, 211–224. https://doi.org/10.1016/j.jmst.2021.07.036

Yelamanchi, B., MacDonald, E., Gonzalez-Canche, N. G., Carrillo, J. G., & Cortes, P. (2020). The mechanical properties of fiber metal laminates based on 3D printed composites. *Materials*, *13*(22), 5264. https://doi.org/10.3390/ma13225264

Yoozbashizadeh, M., Yavari, P., & Khoshnevis, B. (2018). Novel method for additive manufacturing of metal-matrix composite by thermal decomposition of salts. *Additive Manufacturing*, *24*, 173–182. https://doi.org/10.1016/j.addma.2018.09.029

Zhang, K., Chen, Y., Weng, L., Cheng, S., Yu, S., & Zeng, T. (2024). Fabrication of dense Cf/SiC ceramic matrix composites using 3D printing and PIP with short carbon fibre as a toughening material. *Virtual and Physical Prototyping*, *19*(1), e2425381. https://doi.org/10.1080/17452759.2024.2425381

Zhou, L., Miller, J., Vezza, J., Mayster, M., Raffay, M., Justice, Q., ... & Bernat, J. (2024). Additive manufacturing: A comprehensive review. *Sensors*, *24*(9), 2668. https://doi.org/10.3390/s24092668

11 Life Cycle Assessment of Additive Manufactured Materials

Mohd Imran Ansari, Ritesh G Mahajan, Sachin S Chavan, Prachi A Dixit, and Mahesh A Naik

11.1 INTRODUCTION

11.1.1 The Evolution of Manufacturing and Its Environmental Toll

For many years, industrial production has been directed by established manufacturing methods, especially subtractive procedures, for example, computer numerical control (CNC) machining processes. To generate a finished product, these techniques entail removing materials from the solid block by drilling, cutting, and milling. While effective for mass production, subtractive techniques generate staggering material waste. Over 7.6 billion tons of discarded manufacturing is produced worldwide each year, which supplies landfill excess and supply depletion. In addition to trash, subtractive operations need a lot of energy and account for 23% of the world's CO_2 emissions. The environmental harm includes deforestation for raw material exploration, water pollution from machining coolants, and air pollution from particulate emissions. Alternative demands are the observation of the circular economy's three guiding principles: reuse, reduce, and recycle. These principles should be increased to avoid following an unsustainable path (Fredriksson, 2019; Peng et al., 2017; Peng et al., 2018).

11.1.2 Additive Manufacturing: A Paradigm Shift

Additive manufacturing (AM), or 3D printing, emerged in the 1980s as a disruptive alternative to traditional manufacturing. AM builds objects layer by layer using digital models, minimizing material waste as subtractive methods produce more waste. AM innovations like stereolithography (SLA) pave the way for modern processes such as fused deposition modeling (FDM) and laser additive manufacturing (LAM). AM's design freedom allows one to make complicated shapes like lattice structures and interior channels that CNC machining cannot. For example, Airbus's 3D-printed cabin brackets reduced its weight by 30%, and lowered fuel consumption and emissions. AM also promotes local production, and reduces transport costs and emissions from the supply chain. Industries like aerospace and healthcare are leading the charge,

DOI: 10.1201/9781003643531-11

while small and medium enterprises (SMEs) have not been able to adopt the system due to high upfront costs (Kellens et al., 2017; Niaki et al., 2019).

11.1.3 The Promise and Perils of AM Sustainability

AM can potentially decrease material wastage by 70–90% in aeronautics and the automotive industry, as shown in studies like life cycle assessments (LCAs). GE Aviation presented an example where 3D-printed fuel injectors have reduced their weight by 25% and increased fuel efficiency by 15% by integrating 20 components into a single part. Excess energy consumption processes such as laser sintering in LAM use two to three times more energy than CNC machining, and SLA's ultraviolet (UV) curing and toxic resin disposal methods pose environmental risks. Material selection is very complicated from a sustainability point of view as bio-based polylactic acid (PLA) resins have a 70% lower carbon footprint than petroleum-based acrylonitrile butadiene styrene (ABS), but lacks durability. These compromises underline the need for a holistic assessment that balances waste reduction with energy consumption, toxicity, and material circularity. (Arrizubieta et al., 2020).

11.2 ADDITIVE MANUFACTURING TECHNOLOGIES AND ENVIRONMENTAL IMPACT

11.2.1 Technical Overview of Key AM Processes

Additive manufacturing refers to a group of technologies that manufacture objects through the layer-by-layer deposition of material, unlike subtractive methods such as CNC machining. FDM, SLA, and LAM are three of the most used AM processes, which offer distinct workflows, applications, and environmental effects (Figure 11.1), with different forms of material compatibility, energy demand, achievable resolution, and consumable requirements. Recent studies show that the potential of AM to reduce material wastage and manufacture complex geometries must be considered critically for long-term sustainability and practical integration into manufacturing ecosystems, but taking into account the energy consumption and consumables associated with AM (Cosmi & Dal Maso, 2020; Daminabo et al., 2020; Lyons, 2012; Mercado-Colmenero et al., 2019; Moradi et al., 2021; Serhan et al., 2020; Studart, 2016; Ulkir et al., 2022; Vock et al., 2019; Wang et al., 2022).

Fused deposition modeling

- Workflow: FDM melts and processes parts layer by layer by extruding thermoplastic fibers (e.g. ABS, PLA) through a heated nozzle. The layers bond through thermal fusion and solidify as they cool.
- Applications: Widely used in prototyping, tooling, and low-cost production (e.g. automotive dashboards, consumer goods). Startups like ICON use FDM for 3D-printed homes, using concrete composite filaments to build structures in less than 24 hours.
- Strengths: Low machine costs (USD 200–5000), material versatility, and minimum post-processing required.

- Limitations: Poor surface finish, anisotropic strength and emissions of ultrafine particles during printing.

Stereolithography

- Workflow: SLA builds parts bottom-up by selectively curing (hardening) liquid photopolymer resin in a vat with a UV laser. Supports are removed manually, and parts may be UV cured for final hardening.
- Applications: Mostly used for high-precision dental implants, jewelry making, and preparing injection molds. For example, Formlabs' dental resins enable the chairside production of crowns with 25 μm accuracy.
- Strengths: Exceptional resolution (<50 μm layer thickness), smooth surface finish.
- Limitations: Toxic resin disposal (only 20% is recycled globally), energy-intensive post-processing (isopropyl alcohol (IPA) baths, UV ovens), and limited material options.

Laser additive manufacturing

- Workflow: Laser metal AM is also referred to as powder bed fusion (PBF) and works by using lasers to sinter or melt metal powders (e.g. titanium, aluminum) layer by layer in an inert gas argon/nitrogen atmosphere.
- Applications: Aerospace components (e.g. GE Aviation's fuel nozzles), medical implants (e.g. hip stems), and lightweight parts for the automotive industry. Siemens Energy uses LAM to produce gas turbine blades with internal cooling channels, reducing weight by 30%.
- Strengths: High strength-to-weight ratios, design freedom for complex geometries.
- Limitations: High energy consumption (100–150 MJ/kg), costly metal powders (USD 300/kg for titanium), and unused powder waste (40–50% discarded per build).

11.2.2 Comparative Life Cycle Assessment: AM vs. CNC Machining

In order to quantify the environmental benefits and disadvantages of AM, we compared it with CNC machines on three metrics: material waste, energy consumption, and carbon footprint.

Material waste

- CNC machining: Subtractive processes often discard 50–90% of the raw material. For example, the machining of a titanium aircraft bracket from a billet produces 12 kg of waste per 1.5 kg of finished product.
- AM: Additive processes only use the material required, plus supporting equipment. Waste ranges from 5% (LAM) to 20% (FDM). Airbus reported a 95% material saving using AM compared to CNC on the 3D-printing of cabin brackets.

Energy consumption

- CNC machining: Energy use averages 50–80 MJ/kg, driven by spindle motors and coolant systems.
- AM: Energy demands vary widely:
  - FDM: 10–30 MJ/kg (low-temperature extrusion)
  - SLA: 40–60 MJ/kg (UV lasers and post-curing)
  - LAM: 100–150 MJ per kg (laser sintering, inert gas systems). A study from the Massachusetts Institute of Technology (MIT) in 2023 found that, although AM reduces waste, its higher energy consumption may offset the carbon savings if it is not powered by electricity.

Carbon footprint

- CNC machining: Emits 5–8 kg CO_2/kg of titanium parts (primarily from material waste and machining energy)
- AM: Emissions depend on energy sources:
  - Grid-powered LAM: 12–15 kg CO_2/kg
  - Renewable-powered LAM: 3–5 kg CO_2/kg

For context, GE Aviation's 3D-printed LEAP engine nozzles reduced life cycle emissions by 240,000 tons annually due to fuel efficiency gains.

11.2.3 Environmental Challenges in AM

Despite its promise, AM faces significant sustainability challenges linked to raw materials, and post-processing.

Material toxicity

- Polymers: ABS fibers emit styrene (a carcinogen) during printing and pose a risk to the respiratory tract when ultrafine particles are used. Although PLA is biodegradable, it releases microplastics into the waste stream.
- Resins: SLA resins contain dangerous monomers such as acrylates. Less than 30% of them are recycled because of cross-linking after treatment.
- Metals: LAM produces nanoparticles (e.g. titanium dioxide) that are associated with pulmonary inflammation in the workplace.

Energy intensity

- High energy consumption processes such as LAM account for 60–80% of the environmental footprint of AM. For example, 1 kg of titanium requires 1.2 MWh to be melted. This is equivalent to the energy supply of a six weeks in an American household.

Mitigation strategies:

- Renewable integration: WASP's solar-powered 3D printers can cut energy costs by 70%.
- Process optimization: The AI integration and optimization of parameters can reduce LAM energy use by 25%.

Post-processing waste

- SLA: It requires IPA baths to remove uncured resin that leads to the generation of 5–10% of toxic waste per kg of parts.
- LAM: Unused metal powder needs to be filtered and cleaned after each part manufacturing, which leads to reduced contamination levels of > 0.1% by rendering it unusable.

11.2.4 Case Study: Aerospace Component Manufacturing

GE Aviation's catalyst turboprop engine

- Technology: LAM (PBF)
- Outcomes:
  - Consolidated 855 parts into 12, reducing weight by 30%.
  - Cut fuel consumption by 15%, saving 1.4 million liters annually per engine.
- LCA findings:
  - Material waste: Reduced from 90% (CNC) to 5% (LAM)
  - Energy penalty: 20% higher energy use vs. CNC, offset by fuel savings over the engine's life cycle.

11.3 MATERIAL SELECTION AND SUSTAINABILITY

11.3.1 Material Types in AM and Their Environmental Footprints

The selection of materials plays a crucial role in enhancing the sustainability of AM, as the choice has a direct impact on waste generation, energy consumption, and toxicity production. Three categories of primary materials have been analyzed: polymers, metals, and resins (Chen et al., 2021; Cojocaru et al., 2022).

Polymers

- PLA: It is produced from a renewable source such as corn starch and has a 70% lower carbon footprint than petroleum-based plastics. It is biodegradable under industrial conditions of composting (50–60°C), but in landfills it is degraded to a minor extent and breaks down into microplastics. The tensile strength of PLA (50–70 MPa) is sufficient for prototyping purposes, but is insufficient for application under load.

- ABS: As an oil-based thermoplastic, ABS offers higher tensile strength (40–50 MPa) and resistance to heat (80–100°C). However, its production emits 3.5 kg of CO_2 per kilogram of material, and while printing, it releases styrene, a carcinogen, at a concentration of 0.1–1.2 mg per mg of material, which is five times the safety limit set by the Occupational Safety and Health Administration.
- Recycled polyethylene terephthalate glycol (PETG): Recycled PETG fibers (for example, Refil) have emerged as a closed-loop alternative, reducing virgin plastic consumption by 90%. However, repeated recycling cycles reduce the mechanical properties by 20–30%.

Metals

- Titanium (Ti-6Al-4V): Titanium is widely used in the aerospace industry as it offers a high tensile strength (830 MPa), but requires high energy to produce a part (200–250 MJ per kg). The by-product titanium dioxide used for recycling saves 95% of virgin production, but it has some contamination risks (> 0.1% of oxygen rendering is unusable).
- Aluminum (AlSi10Mg): The light weight and corrosion resistance of aluminum requires 95% more recycling than titanium material. However, it has a lower melting point, i.e. 570°C, compared to 1668°C for titanium.
- Stainless steel (316L): Stainless steel is an ideal material for medical implants because it is durable and resistant to corrosion. Its manufacture emits 6.5 kg of CO_2 per kilogram of material, but the reuse rate of AM powder is 80%, with proper sieving.

Resins

- Standard UV resins: These resins, which are composed of acrylates and photo-initiators, release volatile organic compounds (VOCs) during printing. Less than 20% of them are recycled because of cross-linking after treatment.
- Bio-based resins: Innovations like Algix's algae-based resins reduce toxicity and carbon emissions by up to 50%. However, they cost 30% more and lack the mechanical stability of traditional resins.

11.3.2 Eco-Design Strategies

Eco-design integrates sustainability into AM processes through material efficiency and functional optimization.

- Topology optimization: AI-driven software (such as Autodesk Generative Design) removes non-critical sections from actual components, which reduces their weight by 40% without compromising their robustness. For example, Airbus redesigned the A320's partitions with topological optimization that cut the use of materials by 45%, resulting in the saving of 465 tons of CO_2 per year.
- Lattice structures: The inner framework of the lattice mimics natural cell structures and improves the strength-to-weight ratio. The National Aeronautics and Space Administration (NASA)'s 3D-printed Mars habitat components use gyro-lattices to reduce weight by 60%, while preserving structural integrity. Similarly, Adidas 4D-FWD running shoes use flexible frames to adjust the cushioning, which reduces the midsole material by 30%.
- Lightweight production: Consolidating several components into single 3D-printed parts reduces both the weight of the part and the energy consumption. For example, Porsche's 3D-printed pistons reduce its weight by 10% and increase engine efficiency by 30%.
- Material efficiency: AM's almost flat-topped production minimizes machine waste. For example, GE Additive has reduced material waste from 90% (CNC) to 5% (AM) in the production of turbine blades, saving 15 tons of titanium per year.

11.3.3 Recycling Challenges and Innovations

Despite AM's potential, material circularity remains a hurdle.

- Polymer degradation: Recycled thermoplastics, such as ABS, lose 20–30% tensile strength after one cycle due to the degradation of the polymer chain. Closed-loop systems, such as Filabot's filament recycling facilities, mitigate this by mixing virgin and recycled pellets.
- Metal powder contamination: LAM's unused metal powder absorbs moisture and oxygen, which limits its reuse. Siemens has developed a vacuum separation system that reduces contamination to < 0.05% and allows for 95% powder reuse.
- Resin recycling: Cross-linked SLA resins cannot be remelted, but chemical recycling (e.g. solvolysis) can break them up into monomers, which can be re-processed. A French start-up, Carbios, uses enzymatic recycling to extract 90% of the monomers in resins, although the cost remains prohibitive.
- Policy gaps: Only 15% of AM waste is recycled worldwide because of the lack of regulation. The European Union (EU)'s Circular Economy Action Plan calls for 50% recycled content of anti-materials by 2030, but enforcement is still in its infancy.

11.4 ECONOMIC AND SOCIAL DIMENSIONS OF AM

11.4.1 Cost-Benefit Analysis of AM Adoption

The economic viability of AM depends on the balance between high upfront costs and long-term savings.

- Capital investment: Industrial AM systems, such as laser powder bed fusion (LPBF) machines, cost USD 500,000 to 1.5 million, while entry-level FDM printers cost USD 100,000 to 500,000. For SMEs, this gap creates barriers to entry. For example, an automotive supplier based in the USA reported a three-year repayment period after investing USD 1.2 million in LPBF prototyping, while SMEs in developing countries often lack access to finance.
- Operational costs
  - Material costs: Titanium powder costs USD 300 per kg compared to USD 50 per kg for bulk titanium. However, AM's close-to-net shape production reduces the cost of machine tooling. Volkswagen saves USD 160,000 a year by 3D-printing its own jigs and fixtures, which eliminates the cost of CNC programming.
  - Labor costs: AM reduces the manual work involved in post-treatment, but requires qualified technicians. GE Additive trains engineers in AM design and simulation, with an average salary of USD 85,000 per annum.
- Return on investment: Large corporations achieve returns on investment by partly consolidating and lightening their production. Boeing's 3D-printed titanium brackets cut the plane's assembly time by 80%, saving USD 2.5 million. On the other hand, SMEs often have difficulty scaling AM because of limited production volumes.

11.4.2 Social Implications: Labor Markets and Accessibility

AM disrupts traditional manufacturing labor dynamics, while democratizing production.

- Job displacement: Automation in AM is threatening 15% of jobs in CNC manufacturing by 2030. For example, German tool manufacturers reduced the role of CNC operators by 12% after the adoption of AM for molding.
- New opportunities: AM is creating a demand for AM designers, simulation engineers, and powder-spray specialists. The EU plans to create 500,000 new jobs in AM by 2035, and training programs such as Siemens AM Academy is bridging the skills gap.
- Democratization of manufacturing: Small portable 3D printers (< USD 500) are empowering local innovation. In rural India, farmers use FDM printers to produce irrigation components locally, thus reducing the dependence on expensive imports. Similarly, Africa is training communities to print medical supplies using 3D printers, cutting the lead time from six months to just 48 hours.

- Ethical concerns
  - Intellectual property: Digital file-sharing risks and counterfeiting present challenges. Nike filed 23 patent lawsuits in 2022 related to 3D-printed sneaker designs.
  - Quality control: Unregulated alternative medical devices (e.g. prostheses) pose a risk to patient safety. The US Food and Drug Administration (FDA) Guidelines of 2023 require the strict certification of 3D-printed medical devices.

11.4.3 Automotive: Porsche's 3D-Printed Pistons

- Porsche used LAM to manufacture bionic design pistons for its sports car, the 911 GT2 RS, which demonstrated the role of AM in high-performance automotive engineering.
- Design: Generative algorithms have designed a topologically optimized structure with internal cooling ducts, which has reduced the weight by 10% (from 612 to 550 g).
- Performance: Lightweight pistons have increased the engine's efficiency by 30%, giving an acceleration boost of 0 to 60 mph (2.7 vs. 3.0 seconds).
- Cost-benefit: Although each piston costs USD 2000 compared to USD 400 for CNC-machined models, the performance improvements justify the price premium for the limited-edition models.

Post-processing tasks include heat treatment with stress relief (650°C for four hours) and accurate machining of the combustion surfaces.

11.4.4 Medical: Stryker's 3D-Printed Spinal Cages

Stryker's Titanium TL Curved Posterior Lumbar Cage, manufactured via electron beam melting (EBM), highlights AM's impact on personalized healthcare.

- Design: Patient-specific porous structures mimic the trabecular bone, promoting 40% faster osseointegration (bone growth) compared to traditional lumbar cages.
- Sustainability: On-demand production reduces material waste by 50% and inventory costs by USD 1.2 million annually across Stryker's facilities.
- Clinical outcomes: A 2023 study of 200 patients showed 30% shorter recovery times and 95% implant success rates.

Regulatory hurdles include FDA approval for pore size (300–800 μm) and the sterility validation of EBM titanium powder (Li et al., 2015).

11.5 FUTURE RESEARCH DIRECTIONS

11.5.1 Renewable Energy Integration

The energy intensity of AM remains a critical barrier to sustainability. Future research must prioritize integrating renewable energy sources (e.g. solar and wind) into AM systems. For example, MIT's solar sinter project uses concentrated solar energy to power selective laser sintering (SLS) machines, reducing energy costs by 70% in desert environments. Similarly, Siemens Energy is developing wind-powered LAM systems for offshore turbine component production. Challenges include stabilizing energy supply for continuous operation and retrofitting existing AM machines. Hybrid systems that combine grid power and renewable electricity (e.g. Tesla Powerwall-backed 3D printers) could bridge the gaps during low sunlight or wind periods.

11.5.2 Closed-loop Material Recycling

Achieving circularity in AM requires innovations in material recovery and reuse. For polymers, enzymatic recycling (e.g. Carbios' PETG breakdown) can recover 90% of monomers from degraded filaments. For metals, Siemens' vacuum-sieving technology reduces titanium powder contamination to < 0.05%, enabling 95% reuse rates. Resins remain problematic due to cross-linking, but solvent-based recycling shows promise in recovering 60–70% of monomers. Future work should standardize recycling protocols and incentivize closed-loop systems through policies like the EU's extended producer responsibility mandates.

11.5.3 Social Life Cycle Assessment Frameworks

Current LCAs neglect social equity metrics such as labor conditions, community health, and accessibility. A 2023 United Nations Environment Program (UNEP) report proposed social life cycle assessment (SLCA) frameworks to quantify AM's impact on:

- Workforce safety: Nanoparticle exposure limits for LAM operators.
- Democratization: Accessibility of AM hubs in rural/developing regions.
- Education: Training programs for underrepresented groups.

For instance, South Africa's 3D Printing for Good initiative tracks how AM adoption affects local employment and healthcare access. Future studies should adopt these frameworks to ensure AM growth aligns with the United Nations' Sustainable Development Goals (SDGs) (Din, 2006; Klöpffer & Grahl, 2009; Schaltegger, 1996; Suh et al., 2016).

11.6 MATERIALS AND METHODS

11.6.1 Methodology Overview

This study adopts a comparative LCA methodology to assess the environmental impact of AM technologies—namely FDM, SLA, and LAM—against traditional subtractive methods such as CNC fabrication. The assessment covers three main metrics: waste of materials, energy consumption, and carbon emissions.

11.6.2 Data Sources

Data has been collected from peer-reviewed academic journals, industry white papers (e.g. GE Aviation, Siemens), and sustainability databases such as the International Energy Agency (IEA), the Intergovernmental Panel on Climate Change (IPCC) and the World Economic Forum (WEF)'s reports for the period 2022–2024. Case-specific values, such as emission or waste rates, have been verified by means of manufacturers' specifications and audits carried out by third parties.

11.6.3 AM Process Simulation

Each AM technology has been simulated using Autodesk NetFab and ANSYS Additive Suite for the modeling of material deposition, heat input, and post-treatment requirements. The SLA and LAM processes included stages to support removal, curing, and the recovery of dust.

11.6.4 LCA Modeling

The LCA was based on ISO 14040 and 14044 standards. The impact assessment was carried out using SimPro software, which integrates the values of the EcoValue database for materials (e.g. ABS, PLA, Ti-6Al-4V). Emissions factors have been adjusted according to the energy mix (renewable vs. grid-based power). Figure 11.1 illustrates the LCA framework (Din, 2006).

11.6.5 Test Objects

Standardized test subjects (e.g. ASTM D638 tensile samples and turbine blade models) have been designed by FDM, SLA, and LAM. The materials used were PLA,

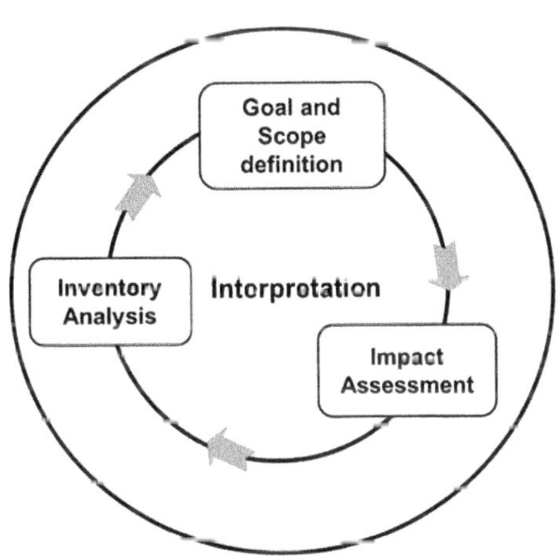

FIGURE 11.1 Life Cycle Assessment Framework.

ABS and Ti-6Al-4V. The dimensions and density of the foundations were kept constant during the test runs to ensure comparability.

11.6.6 Energy and Waste Measurement

Meters recorded the total energy consumption for each printing cycle. During post-production, digital scales measured aid and waste materials. SLA waste was calculated from the loss of uncured resin and IPA solvent, while LAM waste was based on unprocessed, unused powder.

11.6.7 Emissions Testing

In FDM and SLA, the VOC and ultrafine particulate matter emissions were measured with the particulate matter (PM) sensor in the Air Assure device located 1 m away from the printshop. In the LAM project, nanoparticles were sampled from airborne particulates using nanoparticles' aerosol monitoring equipment under inert gas conditions.

11.7 RESULTS AND DISCUSSION

11.7.1 Material Waste

AM processes demonstrated significantly lower material waste compared to CNC:

- CNC machining: 50–90% waste
- FDM: ~20%
- SLA: ~10
- LAM: ~5% (but 40–50% powder recycling loss per build)

11.7.2 Energy Use

LAM exhibited the highest energy demand, driven by laser power and inert gas circulation:

- FDM: 10–30 MJ/kg
- SLA: 40–60 MJ/kg
- LAM: 100–150 MJ/kg
- CNC: 50–80 MJ/kg

11.7.3 Carbon Footprint

CO_2 emissions varied depending on the energy source:

- CNC: 5–8 kg CO_2/kg
- FDM: 2–5 kg CO_2/kg
- SLA: 3–6 kg CO_2/kg

- LAM (grid-powered): 12–15 kg CO_2/kg
- LAM (renewable-powered): 3–5 kg CO_2/kg

11.7.4 Case Study Outcomes

GE's catalyst engine using LAM consolidated 855 parts into 12, reduced weight by 30%, and cut fuel usage by 15%. LCA showed 240,000 tons of CO_2 avoided per year due to AM integration.

11.7.5 Outcome of the Results

The GE catalyst engine using LAM has reduced 855 parts to 12, reduced the mass by 30%, and reduced fuel consumption by 15%. LCA has shown that 240,000 tons of CO_2 is avoided annually through AM integration.

11.7.6 Material Waste and Resource Efficiency

- Significant reduction of waste: AM technologies such as LAM and SLA show a material waste rate of 5–10%, compared to 50–90% in traditional CNC processes. In high-value metals such as titanium, the reduction of waste is reflected in cost savings and improved resource efficiency. The ability to produce parts of almost circular shape means that post-processing and finalization operations (which usually generate waste) are minimized.
- Impact on raw material conservation: For aircraft components, the switch from CNC to LAM reduces the per-part material waste from 12 kg to < 1 kg. In the case of polymers, the majority of AM waste is generated by supporting structures, which are increasingly optimized by means of generative design tools.

11.7.7 Energy Consumption and Process Efficiency

- The energy-intensive nature of AM: The study shows that LAM uses 100–150 MJ per kg—almost two to three times more than CNC machining (50–80 MJ per kg). SLA uses 40–60 MJ per kg, largely due to the UV curing and post-treatment process. FDM has the lowest energy consumption of all AM technologies (10–30 MJ per kg), which makes it ideal for low-power applications.
- Offset through operational efficiency: Despite higher energy inputs per unit, the life cycle consumption of applications such as GE engine components is lower due to the energy savings achieved by part-lightweighting. The consolidation of several components into one AM component reduces assembly costs and tooling requirements.

11.7.8 Carbon Emissions and Environmental Footprint

AM carbon output varies significantly by energy source. LAM, which is powered by grid electricity, emits 12–15 kg of CO_2 per kilogram of paper. When renewable energy is used, the same process can reduce emissions to 3–5 kg CO_2 per kg, which

TABLE 11.1
Manufacturing Processes and Their Energy Consumption, Waste Production and CO_2 Emissions

Process	Material waste	Energy use (MJ/kg)	CO_2 emissions (kg CO_2/kg)
CNC	50–90%	50–80	5–8
FDM	~20%	10–30	2–5
SLA	~10%	40–60	3–6
LAM	~5%	100–150	12–15 (grid), 3–5 (renewable)

is equal to or less than CNC emissions (5–8 kg CO_2 per kg). FDM and SLA show intermediate emissions, which vary according to the type of fiber or resins and the post-treatment demand. Table 11.1 shows the energy consumption and CO_2 emissions of the different processes.

- Life cycle environmental benefits: GE Aviation's 3D-printed nozzle reportedly saves 240,000 tons of CO_2 annually by increasing the efficiency of the engine. The lightweight pistons in the Porsche make the engine 30% more fuel-efficient, reducing fuel consumption and associated emissions per km.

11.7.9 Toxicity and Occupational Health Risks

- Airborne emissions and VOCs: FDM emits ultrafine particles and VOCs, including styrene, which are a risk to the health of the airways. SLA resins contain acrylates and photo-immolators, which are toxic when exposed to the skin or inhaled; untreated resin and IPA waste should be handled with care.
- Metal nanoparticle hazards: LAM may release titanium dioxide and other nano-metals, which are associated with lung inflammation and other long-term health effects. Occupational exposure standards for LAM are still insufficiently developed and require the use of closed chambers and filtration systems.

11.7.10 Material Selection and Sustainability Trade-offs

- Polymers: PLA, a bio-based plastic, has a 70% lower carbon footprint than petroleum-based ABS, but has a lower mechanical life. ABS provides structural integrity, but emits carcinogenic compounds during printing.
- Metals: Titanium (Ti-6Al-4V) offers excellent strength-to-weight ratios, but is very energy-intensive to process and print. Aluminum alloys and stainless-steel show promise for recycling, but require strict contamination control to maintain the quality of the printing.
- Recyclability constraints: SLA resins are not recyclable after curing, and polymer degradation limits the reuse of thermoplastic materials such as ABS and PLA, but the reuse of LAM powder is limited by contamination limits

TABLE 11.2
Materials with Their Source, Strength, Recyclability and CO_2 Emissions

Material	Source	CO_2 (kg/kg)	Recyclability	Strength (MPa)
PLA	Bio-based	1.2	Moderate	50–70
ABS	Petroleum	3.5	Low	40–50
Recycled PETG	Recycled	1.0	High	~45
Ti-6Al-4V	Metal	10–15	Moderate	~830
AlSi10Mg	Metal	8.0	High	~320

(> 0.1% of oxygen or moisture content), although vacuum-sieving systems improve the recyclability of the material. Table 11.2 shows the materials and their origins, recyclability, strength and CO_2 emissions.

11.7.11 Process Optimization and Performance Enhancement

- Topology optimization and part consolidation: Using a regenerative design reduces weight by up to 40%, while maintaining performance and significantly reducing energy consumption during the printing process and throughout the life cycle of the product. Consolidation of 20 to 100 parts into a single printed part eliminates the intermediate steps of assembly, bonding, fastening, and checking.
- Performance benefits translate to environmental gains: Airbus cabin brackets and Porsche pistons, manufactured through 3D printing, have a reduced system weight, leading to fuel efficiency and emission reductions during use.

11.7.12 Socioeconomic Considerations and Scalability

- Democratization vs. industrial exclusivity: While portable FDM printers (< USD 500) support local innovation (e.g. rural Indian farmers and African health centers), industrial AM systems remain prohibitively expensive (> USD 1 million). This hampers the take-up of AM by SMEs. To bridge this gap, government incentives (such as Horizon Europe grants and the US Inflation-Reinforcing Tax Act) are needed.
- Job displacement and skill transformation: AM will be increasing up to 15% of manufacturing jobs that are to be displaced by 2030. It also creates a demand for qualified roles in AM design, simulation, and dust management, which highlights the need for retraining programs.

11.8 CONCLUSION

Additive manufacturing has the potential to be a sustainable innovation to address the environmental, economic, and social issues of traditional manufacturing. This chapter systematically assessed the environmental benefits of AM, while outlining the remaining barriers, and highlighting the need for a comprehensive framework to guide its implementation.

AM is 70–90% more efficient with materials. Case studies in GE Aviation's 3D-printed fuel injectors and Airbus's lightweight cabin fittings illustrate how AM can integrate components and minimize weight, with potential energy savings. However, these savings are negated by the high energy consumption of AM processes (two to three times that of CNC) and the toxicity of the materials used, from ABS styrene emissions to SLA waste in uncured resin. Sustainable alternative fuels, therefore, rely on renewable energy sources, closed-loop recycling and eco-friendly material innovation. It is also important to note that, although materials are a significant factor in the sustainability of AM, another significant factor is the amount of energy consumed during the printing process, especially for thermoplastic-based processes such as FDM. SLA-based printing processes are more energy efficient and have a quicker turnaround time, but the end-life and recycling phases are problematic. This highlights the importance of a full LCA for all processes, from extraction to production to disposal. Moreover, design strategies such as topology optimization, which allow manufacturers to minimize the mass of components without reducing their structural integrity, present additional avenues for sustainability by saving materials and energy. For highly durable materials such as stainless steel, these optimization actions could have a significant direct environmental effect, particularly in metal-based AM processes such as wire and arc AM.

Additive manufacturing holds promise as a more sustainable form of production, particularly for complex, customized, or small-scale production. However, more research, technical refinement, and LCA are needed to fully realize its environmental potential. By expanding the scope of analysis and filling data gaps, AM can become a foundation for sustainable manufacturing both in industry and in research.

REFERENCES

Arrizubieta, J. I., Ukar, O., Ostolaza, M., & Mugica, A. (2020). Study of the environmental implications of using metal powder in additive manufacturing and its handling. *Metals*, 10(2), 261, www.mdpi.com/2075-4701/10/2/261.

Chen, Z., Yang, M., Ji, M., Kuang, X., Qi, H. J., & Wang, T. (2021). Recyclable thermosetting polymers for digital light processing 3D printing. *Materials & Design*, 197, 109189, https://doi.org/10.1016/j.matdes.2020.109189.

Cojocaru, V., Frunzaverde, D., Miclosina, C. O., & Marginean, G. (2022). The influence of the process parameters on the mechanical properties of PLA specimens produced by fused filament fabrication—A review. *Polymers*, 14(5), 886, https://doi.org/10.3390/polym14050886.

Cosmi, F., & Dal Maso, A. (2020). A mechanical characterization of SLA 3D-printed specimens for low-budget applications. *Materials Today: Proceedings*, 32, 194–201, https://doi.org/10.1016/j.matpr.2020.04.602.

Daminabo, S. C., Goel, S., Grammatikos, S. A., Nezhad, H. Y., & Thakur, V. K. (2020). Fused deposition modeling-based additive manufacturing (3D printing): Techniques for polymer material systems. *Materials Today Chemistry*, 16, 100248, https://doi.org/10.1016/j.mtchem.2020.100248.

Din, E. (2006). 14044: Umweltmanagement–Ökobilanz–Anforderungen und Anleitungen. *Deutsche und englische Fassung EN ISO*, 14044, 2006–2010, www.dinmedia.de/de/norm/din-en-iso-14044/325953813.

Fredriksson, C. (2019). Sustainability of metal powder additive manufacturing. *Procedia Manufacturing*, 33, 139–144, https://doi.org/10.1016/j.promfg.2019.04.018.

Kellens, K., Baumers, M., Gutowski, T. G., Flanagan, W., Lifset, R., & Duflou, J. R. (2017). Environmental dimensions of additive manufacturing: Mapping application domains and their environmental implications. *Journal of Industrial Ecology*, 21, 49–68, https://doi.org/10.1111/jiec.12629.

Klöpffer, W., & Grahl, B. (2009). *Ökobilanz (lca): Ein leitfaden für ausbildung und beruf*. John Wiley & Sons, https://onlinelibrary.wiley.com/doi/book/10.1002/9783527627158?msockid=3e61c745a6f36da11f71d109a7686c23.

Li, W., Zhou, J., & Xu, Y. (2015). Study of the in vitro cytotoxicity testing of medical devices. *Biomedical Reports*, 3(5), 617–620, https://pmc.ncbi.nlm.nih.gov/articles/PMC4535150/.

Lyons, B. (2012). Additive manufacturing in aerospace: Examples and research outlook. *Bridge*, 42, 13–19, www.nae.edu/19579/19582/21020/57865/58467/Additive-Manufacturing-in-Aerospace-Examples-and-Research-Outlook.

Mercado-Colmenero, J. M., Rubio-Paramio, M. A., la Rubia-Garcia, M. D., Lozano-Arjona, D., & Martin-Doñate, C. (2019). A numerical and experimental study of the compression uniaxial properties of PLA manufactured with FDM technology based on product specifications. *The International Journal of Advanced Manufacturing Technology*, 103, 1893–1909, https://doi.org/10.1007/s00170-019-03626-0.

Moradi, M., Aminzadeh, A., Rahmatabadi, D., & Hakimi, A. (2021). Experimental investigation on mechanical characterization of 3D printed PLA produced by fused deposition modeling (FDM). *Materials Research Express*, 8(3), 035304, https://iopscience.iop.org/article/10.1088/2053-1591/abe8f3.

Niaki, M. K., Torabi, S. A., & Nonino, F. (2019). Why manufacturers adopt additive manufacturing technologies: The role of sustainability. *Journal of Cleaner Production*, 222, 381–392, https://doi.org/10.1016/j.jclepro.2019.03.019.

Peng, S., Li, T., Wang, X., Dong, M., Liu, Z., Shi, J., & Zhang, H. (2017). Toward a sustainable impeller production: Environmental impact comparison of different impeller manufacturing methods. *Journal of Industrial Ecology*, 21, 216–229, https://doi.org/10.1111/jiec.12628.

Peng, T., Kellens, K., Tang, R., Chen, C., & Chen, G. (2018). Sustainability of additive manufacturing: An overview on its energy demand and environmental impact. *Additive Manufacturing*, 21, 694–704, https://doi.org/10.1016/j.addma.2018.04.022.

Schaltegger, S. (Ed.). (1996). *Life cycle assessment (LCA)—quo vadis?*. Springer Science & Business Media, https://doi.org/10.1007/978-3-0348-9022-9.

Serhan, M., Jackemeyer, D., Long, M., Sprowls, M., Perez, I. D., Maret, W., & Forzani, E. (2020). Total iron measurement in human serum with a novel smartphone-based assay. *IEEE Journal of Translational Engineering in Health and Medicine*, 8, 1–9, https://ieeexplore.ieee.org/abstract/document/9126819.

Studart, A. R. (2016). Additive manufacturing of biologically-inspired materials. *Chemical Society Reviews*, 45(2), 359–376, https://doi.org/10.1039/C5CS00836K.

Suh, S., Leighton, M., Tomar, S., & Chen, C. (2016). Interoperability between ecoinvent ver. 3 and US LCI database: A case study. *The International Journal of Life Cycle Assessment*, 21, 1290–1298, https://doi.org/10.1007/s11367-013-0592-2.

Ulkir, O., Ertugrul, I., Akkus, N., & Ozer, S. (2022). Fabrication and experimental study of micro-gripper with electrothermal actuation by stereolithography method. *Journal of Materials Engineering and Performance*, 31(10), 8148–8159, https://doi.org/10.1007/s11665-022-06875-5.

Vock, S., Klöden, B., Kirchner, A., Weißgärber, T., & Kieback, B. (2019). Powders for powder bed fusion: A review. *Progress in Additive Manufacturing*, 4, 383–397, https://doi.org/10.1007/s40964-019-00078-6.

Wang, M., Li, W., Mille, L. S., Ching, T., Luo, Z., Tang, G., & Zhang, Y. S. (2022). Digital light processing based bioprinting with composable gradients. *Advanced Materials*, 34(1), 2107038, https://doi.org/10.1002/adma.202107038.

12 Impact of Process Parameters on Tribological Properties of Additive Manufactured Polyamide

Prashant Kumar Singh, Sanjay Yadav, Surender Kumar, Gagandeep Roop Rai, and Ahmed Fakhrudin Safdari

12.1 INTRODUCTION

Polyamide (PA) has been regarded as an engineering polymer for decades. Its superior mechanical behavior and high-quality tribological features have enabled the expansion of its use in various industry fields, including the automotive and aerospace industries, and for machinery parts (Alzyod & Ficzere, 2023; Kohutiar et al., 2025; Zhang et al., 2023). Although other plastic materials, such as polylactic acid (PLA) and acrylonitrile butadiene styrene (ABS) have frequently been used in the fabrication of fused deposition modeling (FDM)-based parts, PA offers certain specific advantages, including superior properties such as better tensile strength and higher chemical resistance. PA also offers better flexibility and higher durability in comparison to other plastic materials. The growth in additive manufacturing (AM) technology, especially FDM, has transformed prototyping, as well as manufacturing, with an unprecedented freedom of design, low material waste, and the use of multi-material and complex geometric designs (Cano-Vicent et al., 2021; Daminabo et al., 2020; Ngo et al., 2018). Yet, the manufacture of high-performance polymers such as PA via FDM require a different approach, as the mechanical characteristics, along with the surface quality of fabricated parts, are influenced by the object's mode of fabrication, since one layer over the other, and numerous printing parameters can be configured (Wang et al., 2024; Zhang et al., 2020). As a result, the tribological and mechanical response of FDM produced PA elements is still an unwarrantedly unexplored topic, causing a significant gap between the characteristics of the material and the achievable performance in the form of AM products. Consequently, this chapter presents a methodical evaluation on how major FDM printing variables (nozzle temperature and layer thickness, along with printing speed) affect the mechanical behavior and wear rate of the developed PA components. The results indicate that

DOI: 10.1201/9781003643531-12

increasing the nozzle temperature promotes inter-layer diffusion and bonding, thus improving tensile strength and reducing wear, while the thickness of the layers and speed of printing have a negative effect on these properties, possibly related to lower inter-layer adhesion and a higher number of structural defects (Man et al., 2021; Vidakis et al., 2022). By clarifying such important parameter property relationships, the present study provides valuable information on the FDM process to produce strong and more reliable PA parts.

12.2 METHODOLOGY

The PA filament used in this study was dried at a temperature of 45°C before printing. The part to be printed was designed in SolidWorks software and then imported into the FDM printer (at GO3D, India). The variables used for the manufacture of PA parts are presented in Table 12.1. Among these parameters, some were varied, whereas others were fixed, as provided in Table 12.1.

The computation of tensile strength for the fabricated PA parts was carried out on a universal testing machine (UTM). A pin-on-disc arrangement was utilized to carry out the wear test of the PA samples. The samples were weighed before and after each experimental run. The following formula was used to determine the wear rate of the developed PA samples (Singh et al., 2018):

$$W = \frac{V}{P * L}$$

Here, W is the wear rate (mm^3 per N-m), V is the volume of material lost due to wear (mm^3), P indicates the load applied on the sample (N) and L represents the sliding distance (m). A load of 10 N was applied during each experimental run and each sample covered a sliding distance of 1000 m. The ratio of mass loss due to wear to the density of PA provided the wear volume.

Nine experimental runs were conducted by varying the levels of nozzle temperature, layer thickness, and printing speed. These experimental runs are provided in Table 12.2.

TABLE 12.1
Different Parameters Along with the Levels

Parameter	Level
Nozzle temperature (°C)	240, 250, 260
Layer thickness (mm)	0.1, .02, 0.3
Printing speed (mm/s)	60, 70, 80
Nozzle diameter (mm)	0.4
Platform temperature (°C)	65
Infill density (%)	100

TABLE 12.2
Experimental Runs

Exp. No.	Parameter	Tensile strength (MPa)	Wear rate (mm^3/N-m)
	Nozzle temperature (°C)		
1	240	58	30.5
2	250	61	28.3
3	260	65	26.7
	Layer thickness (mm)		
4	0.1	62	28.4
5	0.2	56	29.1
6	0.3	51	30.3
	Printing speed (mm/s)		
7	60	65	30.8
8	70	63	31.1
9	80	60	31.7

12.3 RESULTS AND DISCUSSION

The tensile behavior of PA with respect to the change in nozzle temperature, layer thickness, and printing speed is shown in Figure 12.1, Figure 12.2, and Figure 12.3, respectively.

The tensile behavior of PA improves with an increase in nozzle temperature, as is evident in Figure 12.1. This happens because of the improved layer adhesion and reduced porosity at higher temperature. The tensile property of PA is negatively affected by the increasing layer thickness. This happens due to there being more voids and weaker bonding between the layers at a higher layer thickness. Additionally, increasing the printing speed negatively influences the tensile behavior of the PA because a higher printing speed offers less time for the bonding of the layers.

The wear behavior of the PA in response to a variation in nozzle temperature, layer thickness, and printing speed is shown in Figure 12.4, Figure 12.5, and Figure 12.6, respectively.

It can be clearly observed from Figure 12.4 that increasing the nozzle temperature decreases the wear rate of FDM-printed PA. This happens because a higher temperature improves the bonding of the material and reduces porosity. Wear behavior becomes more dominant while increasing the layer thickness, as is evident from Figure 12.5. This happens because thin layers have better adhesion, which resists wear. Figure 12.6 shows that a higher printing speed leads to a higher wear rate. This may have happened because a higher speed reduces crystallinity, which causes lower strength to the part, leading to higher wear.

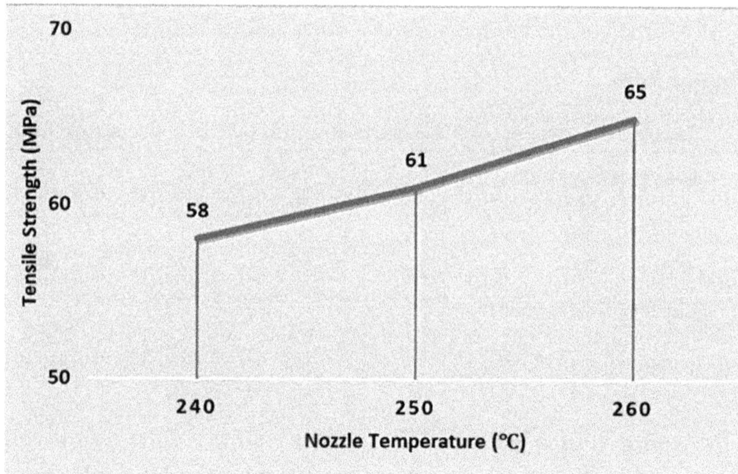

FIGURE 12.1 Effect of nozzle temperature on tensile strength (at a layer thickness of 0.1 mm and a printing speed of 60 mm/s).

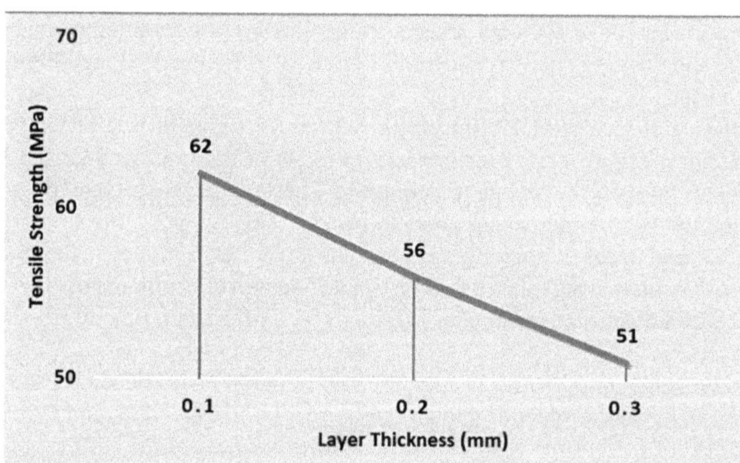

FIGURE 12.2 Layer thickness vs. tensile strength (nozzle temperature = 240°C and printing speed = 60 mm/s).

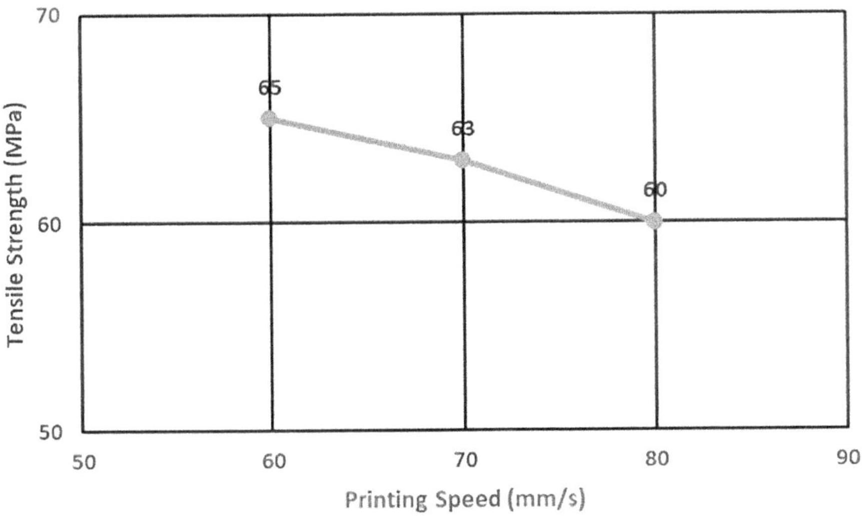

FIGURE 12.3 Impact of variation in printing speed on tensile strength (nozzle temperature = 240°C and layer thickness = 0.1 mm).

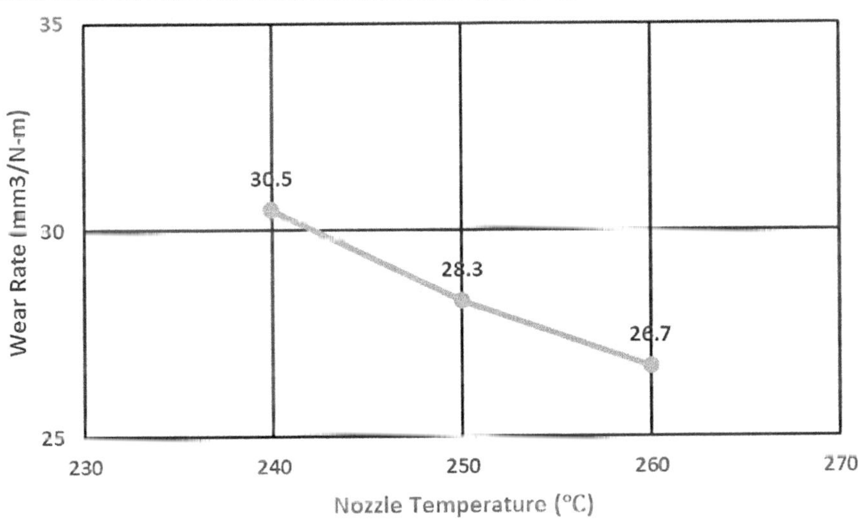

FIGURE 12.4 Effect of nozzle temperature on wear rate (layer thickness of 0.1 mm and printing speed of 60 mm/s).

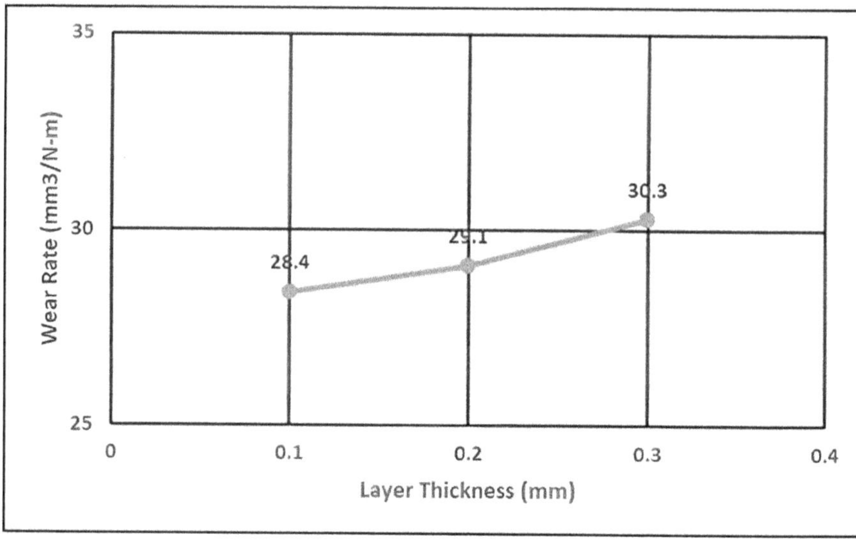

FIGURE 12.5 Impact of variation in layer thickness on wear rate (nozzle temperature = 240°C and printing speed = 60 mm/s).

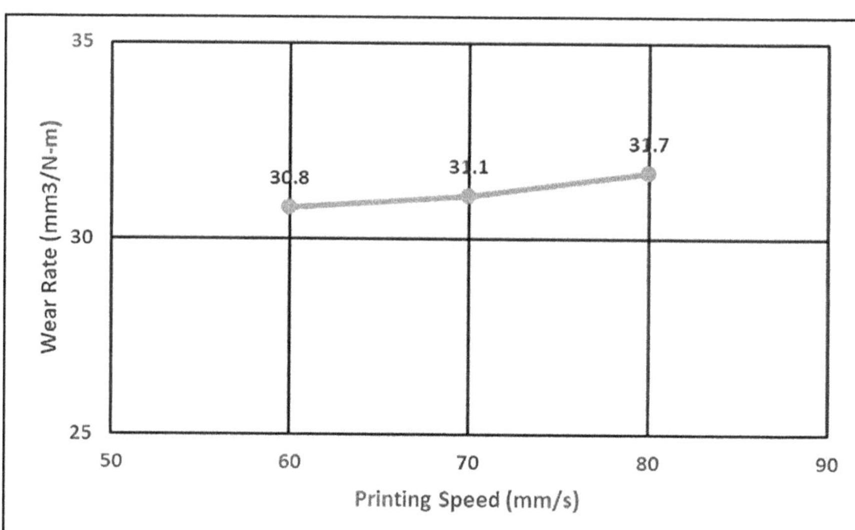

FIGURE 12.6 Impact of variation in printing speed on wear rate (nozzle temperature = 240°C and layer thickness = 0.1 mm).

12.4 CONCLUSION

The current study focuses on investigating the tensile and wear performance of FDM-printed PA material under varying printing conditions. The findings of the study can be concluded as follows:

- PA shows improved tensile strength and a reduced wear rate with increasing nozzle temperature.
- An increment in the layer thickness results in reduced tensile strength, and increases the wear rate of the PA, which indicates that a lower layer thickness should be preferred.
- A higher printing speed also negatively affects both the tensile strength and the wear rate of the PA.

The outcomes of the current study may be utilized in further studies to obtain the optimal printing parameters to improve the mechanical and tribological behavior of FDM-printed PA.

REFERENCES

Alzyod, H., & Ficzere, P. (2023). Optimizing fused filament fabrication process parameters for quality enhancement of PA12 parts using numerical modeling and taguchi method. *Heliyon*, *9*(3), e14445. https://doi.org/10.1016/j.heliyon.2023.e14445

Cano-Vicent, A., Tambuwala, M. M., Hassan, Sk. S., Barh, D., Aljabali, A. A. A., Birkett, M., Arjunan, A., & Serrano-Aroca, Á. (2021). Fused deposition modelling: Current status, methodology, applications and future prospects. *Additive Manufacturing*, *47*, 102378. https://doi.org/10.1016/j.addma.2021.102378

Daminabo, S. C., Goel, S., Grammatikos, S. A., Nezhad, H. Y., & Thakur, V. K. (2020). Fused deposition modeling-based additive manufacturing (3D printing): Techniques for polymer material systems. *Materials Today Chemistry*, *16*, 100248. https://doi.org/10.1016/j.mtchem.2020.100248

Kohutiar, M., Kakošová, L., Krbata, M., Janík, R., Fekiač, J. J., Breznická, A., Eckert, M., Mikuš, P., & Timárová, Ľ. (2025). Comprehensive review: Technological approaches, properties, and applications of pure and reinforced polyamide 6 (PA6) and polyamide 12 (PA12) composite materials. *Polymers*, *17*(4), 442. https://doi.org/10.3390/polym17040442

Man, Z., Wang, H., He, Q., Kim, D.-E., & Chang, L. (2021). Friction and wear behaviour of additively manufactured continuous carbon fibre reinforced PA6 composites. *Composites Part B: Engineering*, *226*, 109332. https://doi.org/10.1016/j.compositesb.2021.109332

Ngo, T. D., Kashani, A., Imbalzano, G., Nguyen, K. T. Q., & Hui, D. (2018). Additive manufacturing (3D printing): A review of materials, methods, applications and challenges. *Composites Part B: Engineering*, *143*, 172–196. https://doi.org/10.1016/j.compositesb.2018.02.012

Singh, P. K., Siddhartha, & Singh, A. K. (2018). An investigation on the thermal and wear behavior of polymer based spur gears. *Tribology International*, *118*, 264–272. https://doi.org/10.1016/j.triboint.2017.10.007

Vidakis, N., David, C., Petousis, M., Sagris, D., Mountakis, N., & Moutsopoulou, A. (2022). The effect of six key process control parameters on the surface roughness, dimensional accuracy, and porosity in material extrusion 3D printing of polylactic acid: Prediction models and optimization supported by robust design analysis. *Advances in Industrial and Manufacturing Engineering*, *5*, 100104. https://doi.org/10.1016/j.aime.2022.100104

Wang, C., He, Y., Lin, Z., Zhao, X., Sun, C., Guo, R., Wang, X., & Zhou, F. (2024). Mechanical and tribological properties of FDM-printed polyamide. *Tribology International*, *191*, 109198. https://doi.org/10.1016/j.triboint.2023.109198

Zhang, X., Fan, W., & Liu, T. (2020). Fused deposition modeling 3D printing of polyamide-based composites and its applications. *Composites Communications*, *21*, 100413. https://doi.org/10.1016/j.coco.2020.100413

Zhang, Y., Wang, C., Yi, Y., Wang, W., & Yang, J. (2023). Synthesis and properties of polyamide 6 random copolymers containing an aromatic imide structure. *Polymers*, *15*(13), 2812. https://doi.org/10.3390/polym15132812

13 Medical Applications of Additive Manufacturing and Future Scope

Reddyvari Loka Vigneswar Reddy, Akant Kumar Singh, Naveen Mani Tripathi, Shailendra Kumar, and Virendra Pratap Singh

13.1 INTRODUCTION

Additive manufacturing (AM), more popularly referred to as 3D printing, is a digital, computer-based manufacturing technology that creates three-dimensional objects layer by layer directly from computer-aided design (CAD) models (Figure 13.1). Compared to conventional subtractive methods like computer numerical control (CNC) machining, AM reduces the wastage of material, and allows the production of complicated geometries and internal features that are difficult or impossible to create using conventional methods (Gibson et al., 2015). The advantages of productive AM include the disruption of the medical sector, where accuracy, customization, and speed are important for patient success (Pérez et al., 2020).

Medical commodity development is highly controlled and time sensitive, with rapid prototyping and testing being essential to meet the requirements of organizations such as the US Food and Drug Administration (FDA). Additive manufacturing has been especially useful in this context by simplifying the rapid repetition of designs, which reduces the time to market of new medical implants and devices (Madrid et al., 2019). Classic applications of AM were restricted to model prototyping, but have now been extended with advancements in processing techniques and materials science to the production of end-use biocompatible components (Picard et al., 2020).

One of the most important uses of AM in the medical field is patient-specific anatomical model customization. Through the translation of magnetic resonance imaging (MRI), computed tomography (CT), or laser scanning into physical 3D models, surgeons and clinicians can produce highly accurate patient anatomy representations (Ventola, 2014). Such models are engaged in pre-surgery planning, implant design, and medical education, minimizing intraoperative risks and enhancing clinical outcomes (Javaid & Haleem, 2017; Rengier et al., 2010).

Additive manufacturing has also transformed the production of prosthetics and implants. Traditional manufacturing processes frequently encountered difficulties in

DOI: 10.1201/9781003643531-13

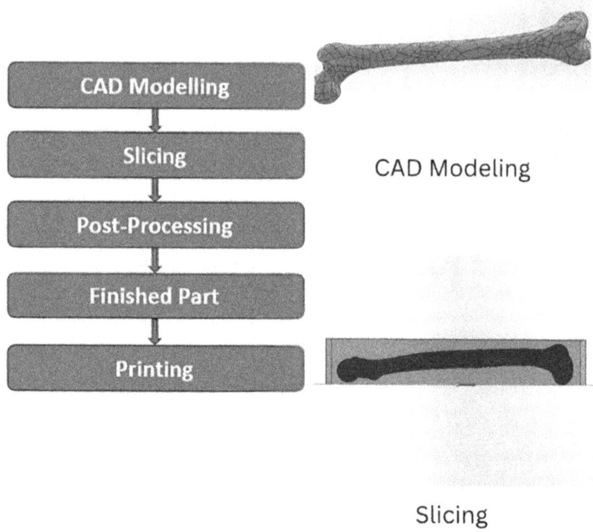

FIGURE 13.1 Overview of the additive manufacturing process.

creating permeable structures that can support bone in growth. Additive manufacturing breaks through this limitation as it makes it possible to produce orthopedic, craniofacial, and dental implants with controlled porosity and intricate geometrics for improved function and biocompatibility (Calignano et al., 2019; Murr et al., 2012). Additive manufacturing has reshaped and revolutionized dental and oral medicine by enabling the fast and accurate fabrication of crowns, aligners, and surgical guides, making production more efficient and patient satisfaction higher (Bhargav et al., 2018; Revilla-Leon et al., 2020).

Another new application is tissue engineering and bioprinting. With the use of bioinks that include living cells, researchers have successfully started printing tissues that contain a functional network of blood vessels (or, in plants, xylem and phloem), deliver nutrients and oxygen, and remove waste, enabling the survival and growth of tissues, cartilage, skin substitutes, and scaffolds. Unfortunately, fully functional organ printing is still a long-term objective. However, recent development shows the ability of AM in regenerative medicine (Murphy & Atala, 2014; Zhou et al., 2020). This technology may eventually solve the problem of a shortage of organ donors.

Moreover, AM has been promising in pharmaceuticals, as well as drug delivery systems. The technology simplifies the creation of tables with customized release mechanisms, and patient-specific dosing. For instance, AM-processed drug delivery systems enable precise geometry and material composition control, enhancing therapeutic efficacy (Trivedi et al., 2018). In addition, during the COVID-19 pandemic, AM's flexibility enabled companies to rapidly produce protective masks, ventilator valves, and diagnostic equipment, to accommodate for shortages in supply, demonstrating its function in emergency medicine.

Additive manufacturing's resilience also spreads into niche areas like cardiology and oncology. In cardiology, stents, valves, and blood vessels are being 3D-printed to increase the accuracy of treatment (Haleem et al., 2018). Oncologists are performing preoperative appraisals and doing radiation planning with 3D growth models, which is enhancing surgical accuracy and patient protection (Thompson et al., 2017).

In spite of these advances, the adoption of AM in medicine has its limitations. Regulatory approval processes remain stringent, biocompatible material substances that interact with a living organism without causing toxic or immunological responses continue to be restricted, and high replicability remains a challenging obstacle for critical implant applications (Montgomery et al., 2020; Vafadar et al., 2021).

The future scope of AM in medicine may be very promising in years to come. The continued investigation of functionally graded materials and intelligent materials is likely to produce implants that more closely resemble the native tissue structure (Gharde et al., 2019). Bioprinting studies are also moving closer to the development of fluffy functional organs, and developments in the multi-material printing of medical devices (Guddati et al., 2019; Melchels, 2012). With decreasing costs and growing material libraries, AM will eventually become an everyday technology across industries like orthopedics, dentistry, and regenerative medicine (Galante et al,, 2019).

13.2 METHODOLOGY

- Research design
- Data collection
- Selection of case studies
- Analytical framework
- Evolution of future scope
- Validation

13.3 RESEARCH DESIGN

13.3.1 Nature of the Research

The research design is qualitative, exploratory, and descriptive in nature. The reason for using a qualitative approach is that it is not concerned with collecting numbers, but with explanation, especially of trends, technological advancements, and the clinical uses of AM in the medical field. The illustrative aspect is due to the evolving nature of AM in healthcare, with new areas of application yet to be identified. The explanatory component enables the study to provide real-life case studies through its findings, and identify related problems (Javaid & Haleem, 2017).

13.3.2 Purpose of the Research Design

The research design was developed to meet the following purposes:

- Chart the development of AM in the medical field, from prototyping to end use
- Compare materials and technologies on the basis of their strength, weakness, and clinical relevance

- Inspect case studies that replicate various branches of the medical field, such as orthopedics, dentistry, cardiology, oncology, and tissue engineering
- Discuss the possible future directions of AM, including controlling environments, smart materials, and bioprinting
- Analyze sustainability limits, including cost effectiveness, material efficiency, and biocompatibility

13.3.3 Case Study-based Approach

The case study approach is the basis of this study. Yin (2014) declares that case studies are suitable for investigating complex, real-world phenomena where the boundaries among technology, process, and outcome are unclear. Each of the case studies, such as AM prosthetics, dental crowns, surgical models, and 3D bio-printed tissues, forms a unit of study. Cases were chosen to present a clinical array and illustrate technological change, providing a balanced summary of AM applications. A multi-case plan was adopted rather than a single case study. This would enhance the rationality of the results by way of cross-comparison.

13.3.4 Research Strategy

The research strategy is composed of six steps, as shown in Figure 13.2:

- Literature survey—perusal of peer-reviewed journals, books, and regulatory literature from 2010–2023

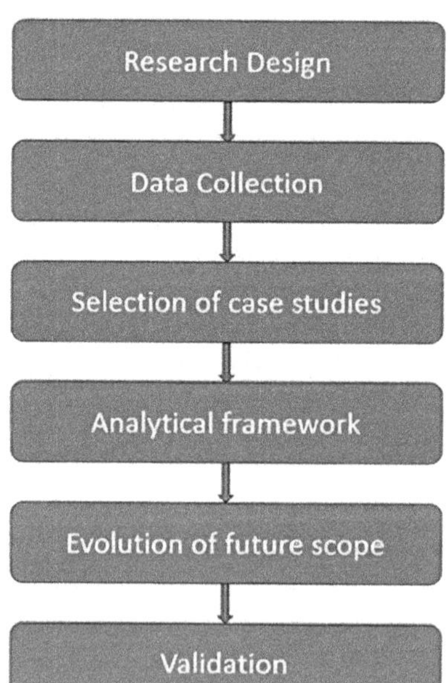

FIGURE 13.2 Research methodology flowchart.

- Database selection—searching through Scopus, PubMed, ScienceDirect, IEEE Xplore, SpringerLink, and Google Scholer
- Case identification—picking medical applications reported in reliable journals or conference proceedings
- Categorization—classifying cases into orthopedics, dentistry, cardiology, oncology, and bioprinting
- Thematic analysis—identifying recurrent benefits, challenges, and innovations across cases
- Validation—checking answers against systematic reviews, FDA standards, and industrial practices

13.3.5 Time Horizon

The research design is cross-sectional. Data collection is done from sources between 2010 and 2023. This period is important because, before 2010, AM in healthcare was principally experimental and attentive on prototyping. Since 2010, there has been an increase in clinical uses, FDA clearances, and the commercialization of AM medical devices (Li et al., 2020).

13.3.6 Data Sources

The research depends completely on secondary sources of data, which are dependable and comprehensive:

- Peer-reviewed journals (ScienceDirect Nature, Springer)
- Conference proceedings recording AM methods
- Regulatory documents (FDA, ISO standards, European Medicines Agency)
- Industrial white papers and case study reports from medical device firms

13.3.7 Justification of Research Design

This design is defended because a quantitative design would not be suitable in light of the unavailability of high-volume numeric datasets in cutting-edge AM medical research. The case study method permits the depth, context, and fusion of technical, clinical, and supervisory understanding. The design offers latitude, allowing investigation into both existing uses and future potential, like organ bioprinting.

13.3.8 Expected Outcomes

The research design is predicted to deliver a snapshot of the role of AM in medicine today, providing comparative knowledge of AM methods and materials. Suggestions are provided for sustainability and the scalability of AM in medical uses. A forward-looking structure is developed to emphasize gaps and opportunities in bioprinting, multi-material printing, and the integration of artificial intelligence (AM) in AM.

13.4 DATA COLLECTION

Information for this study was collected entirely from secondary sources because of the research goals to identify proven and upcoming proposals of AM in the medical field. A methodical review of the literature was led through academic databases, including ResearchGate, IEEE Xplore, SpringerLink, and ScienceDirect, to obtain reliable, peer-reviewed articles. The search used targeted keywords such as "3D printing in medicine," "additive manufacturing in healthcare," "bioprinting," "patient-specific implants," and "medical prosthetics." To identify current developments, the review considered publications from 2010 to 2023, as this period imitates the shift of AM from being used only for prototyping, to clinically cleared medical uses. Sources were journal articles, review papers, books, FDA and ISO regulatory guidelines, and white papers from industry, documenting real-world applications of AM in healthcare. Data gathered included different dimensions, like AM, materials, clinical case studies, and sustainability factors like cost savings, material efficiency, and time-to-market optimization. To justify validity and reliability, data across different sources was cross-checked, and results compared against methodical reviews and medical device regulatory requirements. Although this gave a comprehensive overview, limitations included reliance on secondary data and restricted access to information regarding novel methods like organ bioprinting and multi-material printing.

13.5 SELECTION OF CASE STUDY

The case studies for the research were chosen to offer a picture of the various ways in which AM is utilized in the medical field. The selection was made under three broad criteria: clinical applicability, variety of applications, and access to dependable evidence (Figure 13.3 and Figure 13.4). The case study was mainly selected to illustrate the utilization of AM to enhance medical outcomes, such as implants, prosthetics, surgical guides, or bioprinting.

To capture the variety of AM applications, case studies were taken from several medical areas. In orthopedics, cases included custom hip and knee implants to match patient anatomy. In dentistry, cases like crowns, bridges, aligners, and surgical templates were taken into account because of their broad use and clinical success. Cardiology case studies featuring vascular models and 3D-printed stents were chosen to show the use of AM in pre-surgery planning and treatment growth. Oncology applications were also chosen, specifically tumor models for surgery and radiation planning. Tissue engineering and bioprinting cases were obtained to show the potential future use of AM in regenerative medicine.

Another way to select a case study is by peer-reviewed evidence, regulatory information, or industrial books to support the findings. This proves that selected case studies are supported by trustworthy sources and represented in current practice. They also have future potential. By taking a multiple case study selection approach, the research not only captures the current role of AM in the medical field, but also its broader suggestions for upcoming innovations like organ fabrication and personalized regenerative medicine. This methodology enables cross-case comparison, adding to the overall validity and scope of the study.

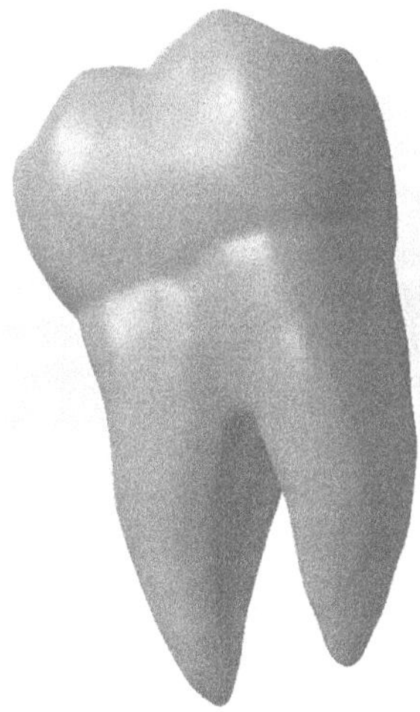

FIGURE 13.3 Dental crown 3D design.

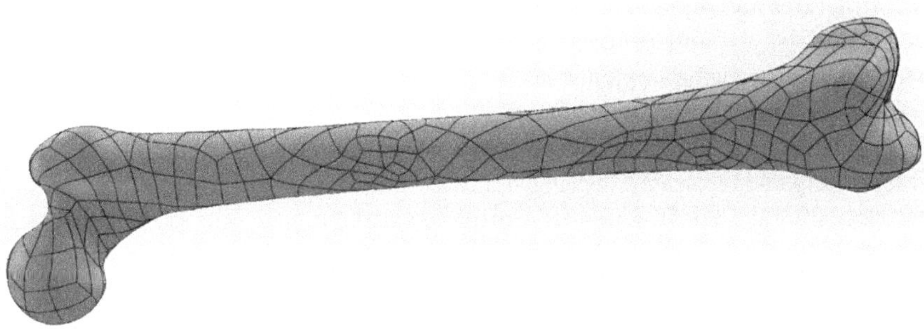

FIGURE 13.4 Orthopedic implant 3D design.

13.6 ANALYTICAL FRAMEWORK

The structure of the study is meant to critically evaluate the input of AM in medical applications using case studies. The structure promotes expressive study, thematic coding, and comparative analysis. Each case study is originally described in depth, paying attention to the category of AM technology employed, materials used, clinical application, and achievements made. This expressive phase confirms the identification of the individual features of every case, giving them a setting for comprehensive analysis.

The cases are categorized into different themes according to medical fields, including orthopedics, dentistry, cardiology, oncology, and tissue engineering.

The study identifies the trends, benefits, and problems of AM. The applications in orthopedics and dentistry focus on exact, accurate and patient-customized fitting, whereas tissue engineering focuses on new methods like bioprinting and support design. These different strategies make it possible to identify and track advantages and regular challenges.

The analytical framework entails a case study comparison and future analysis. The comparison of case studies across various medical areas, and patterns of acceptance refer to different aspects, like behavior, model relatedness, and the acceptance of new things, ideas, or circumstances. The technological maturity indicates how its reliability, readiness, widespread use, and sustainability features are established. The study includes sustainability factors such as material efficiency, avoiding waste, cost-effectiveness, and clinical safety to estimate the overall impact of AM in the medical field. This combined analytical method confirms current uses of AM, but also its possibility for the future, especially in the fields of bioprinting and regenerative medicine.

13.7 EVOLUTION OF FUTURE SCOPE

The future scope of the medical application of AM includes rapid manufacturing and prototyping as a means to become the mainstream technology to develop specific implants, prosthetics, and surgical models. At the start of AM, it was mainly used for rapid manufacturing and prototyping, allowing engineers and doctors to imagine difficult structures and rapid device creation. By improving materials science, as well as AM technology, the area of interest shifted to developing medical devices like dental restorations, orthopedic implants, and anatomical training models. The solidification of AM started find direct use in clinics.

Over the last few years, the development of AM in the medical field has been seen in bioprinting and tissue engineering. The use of bio-inks with living cells has enabled new places for the construction of scaffolds, cartilage, and skin grafts. Research is currently advanced and improving towards printing vascularized tissues and organ-like structures. Although there were small problems and challenges in obtaining fully functional organs, early achievements by researchers in generating growing tissue show the potential of AM in transforming the field from structural applications to regenerative implants and organ replacement, potentially leading to new ways of patient care in the future (Figure 13.5).

The dominant direction for the future of AM is intelligent and multi-material printing. Only a single material is usually used in traditional implants and devices, but now, functionally graded materials can enable implants and prosthetics to replicate the uses and properties of human bone, cartilage, and soft tissue. Titanium polymer composite material and bioresorbable polymers are being investigated to improve biocompatibility, mechanical behavior, and patient safety. The development of technology implies that AM will not only duplicate human biology, but will also reach and interact dynamically with biological systems.

Integration takes place with digital and computerized technologies like AI, machine learning, and cloud-based simulations. AI-based design can correct

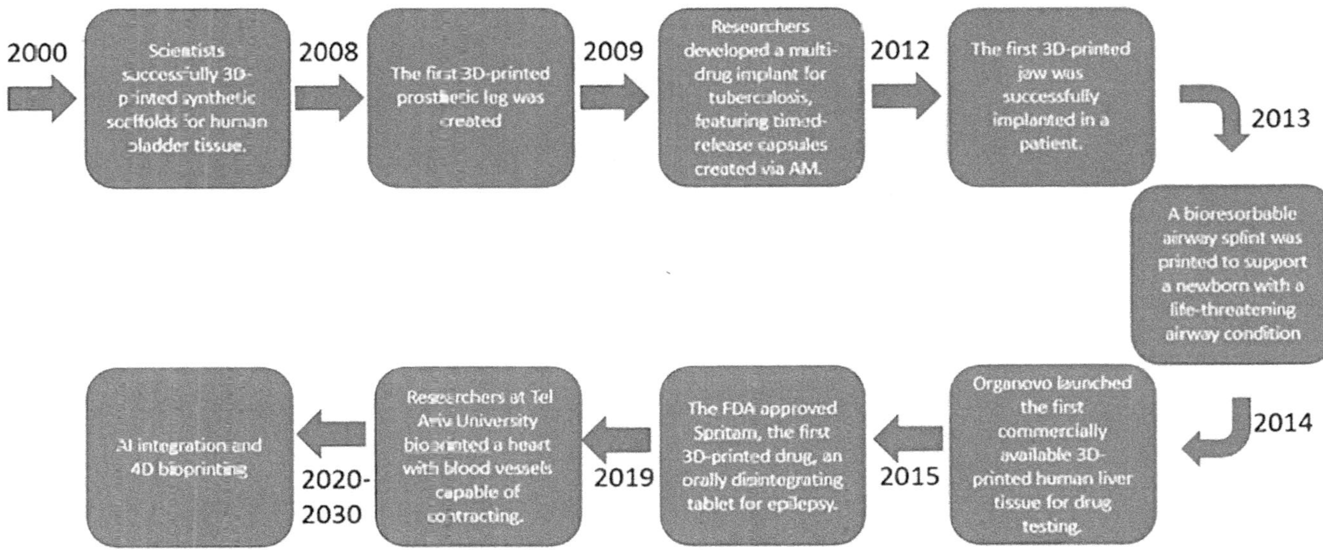

FIGURE 13.5 Evolution of additive manufacturing in medical applications.

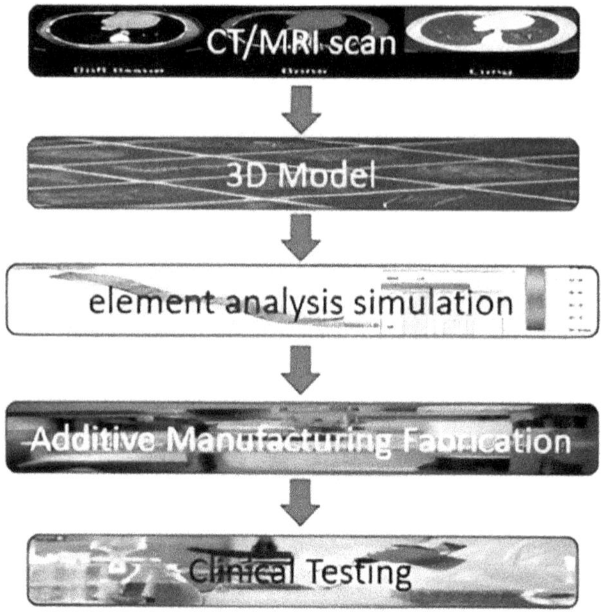

FIGURE 13.6 Validation process of additive manufactured medical devices.

implant dimensions and geometry according to specific patient needs, and real-time monitoring allows for printing defects in sensitive medical devices to be corrected. These improvements can reduce clinical risks, as well as replicability, which can overcome shortcomings in terms of safety and certification by specialists.

The future scope of the development of AM in the medical field will find application in personalized medicine, sustainable production, and access to global healthcare. With reducing costs and improved regulatory environments, the demand for manufactured medical devices in resource-constrained areas will improve. The medical manufacturing highlights of AM are not just about technological advancement, but its development as the main force in achieving equality in world health.

13.8 VALIDATION

The process of validation in AM for medical applications is crucial. This system is aimed at establishing that adapted devices are safe, effective, and compliant for medical use, directly attempting to overcome the limitations of conventional, uniform components (Figure 13.6).

13.8.1 Customization and Accuracy

The starting point in advanced medical scanning is imaging like CT or MRI to obtain a specific patient anatomy. This information is then used to generate a digital 3D model with high precision. The 3D model is very important to design the customized implant so as to provide improved fit over conventional implants and prosthetics. The outcome is a 3D-optimized and reconstructed model with the correct anatomy to maintain high-dimensional accuracy for printing.

13.8.2 Biomechanical Integrity

The validation of the structural capability of an implant involves a thorough finite element modeling or finite element analysis stage. This simulation, which is computer based and analyses the stress distribution on the bone and implant under different loading conditions, allows engineers and surgeons to improve the design so that it can provide mechanical stability. This step guarantees that material properties are beyond failure.

13.8.3 Process and Product Qualification

The AM process itself has been qualified to maintain consistency and reproducibility. The qualification process includes three steps.

The first step is the installation, which entails checking that the equipment needed for AM is properly installed and is consistent with manufacturers' specifications. The second step is the operational requirement. This entails checking that the equipment operates as required to ensure adherence to test plan specifications. The third step of is the performance requirement. This entails checking that the entire manufacturing process can be duplicated constantly under normal working conditions.

13.8.4 Clinical Effectiveness

Normally, validation confirms the clinical advantage, and that the specific patient implants can reduce the difficulty of the surgical step-by-step procedure, reduce the need for interoperative adjustment, and reduce the total recovery time. We can know, for example, by taking the medical application of custom implants in craniomaxillofacial surgery, that it has been verified and will lead to the maximum geometrical accuracy and have less biomechanical properties.

13.9 RESULT AND DISCUSSION

The case studies highlight the developing potential of AM in the medical field. The results indicate that AM has developed well beyond its convectional function of prototyping and has determined the correct method for manufacturing useful and functional special patient-customized medical prosthetics devices. Through all the considered applications (orthopedic, dental, cardiological, and tissue engineering), the combination of AM technologies has resulted in improved accuracy, customization, and efficiency of treatment. These results verify that AM is able to produce a device or manufactured organ with the correct geometrics and internal structures that are impossible to produce through traditional manufacturing processes. In the field of orthopedics, case studies demonstrated that 3D-printed implants and prosthetics constructed from titanium alloys have recorded significant success rates in medical use. For example, specific patient hip and spinal implants have a better fit. They are also reported to require shorter surgeries, and to provide faster recovery post-surgery than normal implants. The capacity to include leaky lattice structures into AM parts has greatly increased bone in growth and osseointegration, resulting in increased implant durability and biocompatibility. In dentistry AM techniques like stereolithography and selective laser sintering have meaningly changed the production of crowns, bridges, aligners, and surgical guides. The accuracy and efficiency of digital processes has enabled dentists to offer same-day refurbishments, improving patient comfort and lowering costs.

The application of 3D-printed functional models in cardiology and surgery have improved preoperative planning and simulation. Surgeons have been able to review specific patients' heart structures or vascular systems to conduct complex procedures, minimizing intraoperative risks and improving surgical accuracy. In the same way, oncology case studies exposed the utility of AM in the production of physical tumor models for surgical and radiation planning, improving the accuracy of tumor resections. In regenerative medicine and tissue engineering, new research has shown the possibility of applying bioprinting to the production of supports, cartilage, and vascularized tissue concepts. Whole organ printing is currently experimental. Recent developments point to an exact direction in functional and useful tissue replacement. The results and discussion also point to the economic and environmental advantages of AM in the medical sector. In contrast to conventional manufacturing, AM techniques save material and energy, and reduce the lead time for personalized products. Manufacturers and hospitals can create specific patient parts on demand, incurring minimized inventory and transportation costs. The analysis recognized a number of issues that need to be overcome for increased acceptance. These range from the high capital cost of equipment to the partial need for biocompatible materials, complex controlling demands, and the need for highly skilled operators. The future combined use of machine learning, AI, and multi-material printing will improve the field going forward. AI is used to improve implant geometry and performance; hybrid materials will enable implants to more closely clone natural tissues. These technologies suggest a future where AM can make bio-printed organs fully functional, make more adaptive prosthetics, and even establish in-hospital production units for personalized medical solutions.

Overall, the findings state clearly and confidently that AM has already made an important contribution to medical product development and patient care (Table 13.1). It is possible to offer accuracy, personalization, and performance. This will keep increasing with advances in materials science and digital addition. The production of medical equipment is also reshaping the entire healthcare ecosystem and its efficiency, directing us towards more stable and specific patient-focused medicine.

TABLE 13.1
Summary of Results of Medical Applications of Additive Manufacturing

Material	Application area	AM used	Findings
Titanium alloys (Ti-6Al-4V)	Orthopedics	Selective laser melting, electron beam melting	Customized implants, improved bone integration
Biocompatible polymers, photopolymers	Cardiology/ surgery	Material jetting, stereolithography	Anatomical heart and vascular models, improved surgical planning, and danger prediction
Dental resin, cobalt chrome	Dentistry	Selective laser sintering, stereolithography	High-accuracy crowns, bridges, and aligners, same-day construction with reduced cost
Hydrogels, collagen, alginate, bio-inks with living cells	Tissue engineering/ bio-printing	Extrusion-based bio-printing	Construction of scaffolds and tissue constructs, progress toward vascularized organs
PLA, PVA, pharmaceutical polymers	Drug delivery/ medical devices	Binder jetting, FDM	Controlled drug release systems and customized surgical instruments developed rapidly
ABS, photopolymer resin	Oncology	Fused deposition modelling, SLA	3D tumor models, aided accurate tumor resections and radiation planning
Smart materials, composite polymers	Future integration	Multi-material and hybrid additive manufacturing	Addition with AI and multi-materials for functional, adaptive implants and organ printing

13.10 CONCLUSIONS

This research clearly indicates that AM is a powerful tool in modern medicine, providing customized and accurate solutions for the care of every patient. Technologies like stereolithography, selective laser sintering, and direct metal laser sintering enable the manufacture of complex geometries and lightweight medical prosthetics that are challenging using conventional manufacturing techniques. In orthopedic and dental uses, AM makes it possible to print 3D implants and prosthetics to suit the specific anatomy of a patient. Titanium implants improve bone integration and reduce the time required for surgery; dental uses such as crowns, bridges, and surgical guides improve accuracy and reduce production time. Anatomical models printed in 3D help surgeons plan for complex operations, reducing intraoperative danger and increasing efficiency. These findings also identify the role of AM in responding to emergencies like the COVID-19 pandemic, where it was used for the rapid manufacture of ventilator components and protective gear.

AM faces some challenges, like the high cost of production, some biocompatible materials, and difficult regulatory approvals. New breakthroughs, like bio-printing, multi-material manufacture, and the addition of AI, are improving the scope of AM. These types of technologies provide the potential for improved material performance, advanced customization, and future uses such as tissue and organ printing. Additive manufacturing is revolutionizing the medical field by personalizing treatments, making them more effective, and making them available to everyone at a lower cost. This ongoing development has the strength to be the future core of medical equipment production and regenerative medicine.

REFERENCES

Bhargav, A., Sanjairaj, V., Rosa, V., Feng, L. W., & Fuh, Y. H. J. (2018). Applications of additive manufacturing in dentistry: A review. *Journal of Biomedical Materials Research Part B: Applied Biomaterials*, 106, 2058–2064. https://pubmed.ncbi.nlm.nih.gov/28736923/

Calignano, F., Galati, M., Iuliano, L., & Minetola, P. (2019). Design of additively manufactured structures for biomedical applications: A review. *Journal of Healthcare Engineering*, 2019, 9748212. https://pmc.ncbi.nlm.nih.gov/articles/PMC6434267/

Galante, R., Figueiredo-Pina, C. G., & Serro, A. P. (2019). Additive manufacturing of ceramics for dental applications: A review. *Dental Materials*, 35, 825–846. www.sciencedirect.com/science/article/abs/pii/S0109564118304263

Gharde, S., Surendren, A., Korde, J. M., Saini, S., Deoray, N., Goud, R., & Kandasubramanian, B. (2019). Recent advances in additive manufacturing of bio-inspired materials. In Prakash C., Singh S., Ramakrishna S., Pabla, B. S., Puri S., & Uddin, M. S. (Eds.), *Biomanufacturing*, 35–68. https://link.springer.com/chapter/10.1007/978-3-030-13951-3_2

Gibson, I., Rosen, D., & Stucker, B. (2015). Design for additive manufacturing. In Gibson, I., Rosen, D., & Stucker, B. (Eds.), *Additive Manufacturing Technologies*. Springer, New York, NY. https://doi.org/10.1007/978-1-4939-2113-3_17

Guddati, S., Kiran, A. S. K., Leavy, M., & Ramakrishna, S. (2019). Recent advancements in additive manufacturing technologies for porous material applications. *International Journal of Advanced Manufacturing Technology*, 105, 193–215. https://link.springer.com/article/10.1007/s00170-019-04116-z

Haleem, A., Javaid, M., & Saxena, A. (2018). Additive manufacturing applications in cardiology: A review. *Egyptian Heart Journal*, 70, 433–441. www.sciencedirect.com/science/article/pii/S1110260818301789

Javaid, M., & Haleem, A. (2017). Additive manufacturing applications in medical cases: A literature-based review. *Alexandria Journal of Medicine*, 53(4), 411–422. www.sciencedirect.com/science/article/pii/S2090506817302282

Li, C., Pisignano, D., Zhao, Y., & Xue, J. (2020). Advances in medical applications of additive manufacturing. *Engineering*, 6(11), 1222–1231. www.sciencedirect.com/science/article/pii/S2095809920302447

Madrid, A. P. M., Vrech, S. M., Sanchez, M. A., & Rodriguez, A. P. (2019). Advances in additive manufacturing for bone tissue engineering scaffolds. *Materials Science and Engineering: C*, 100, 631–644. www.sciencedirect.com/science/article/pii/S0928493118307938

Melchels, F. P. W. (2012). Additive manufacturing of Tissue and Organs. *Progress in polymer Science*, 37(8), 1079–1104. https://doi.org/10.1016/j.progpolymsci.2011.11.007

Montgomery, S. M., Kuang, X., Armstrong, C. D., & Qi, H. J. (2020). Recent advances in additive manufacturing of active mechanical metamaterials. *Current Opinion in Solid State and Materials Science*, 24(5), 100869. www.sciencedirect.com/science/article/pii/S135902862030067X

Murphy, S., & Atala, A. (2014). 3D bioprinting of tissues and organs. *Nat Biotechnol*, 32, 773–785. https://doi.org/10.1038/nbt.2958

Murr, L. E., Gaytan, S. M., Martinez, E., Medina, F., Wicker, R. B. (2012). Next generation orthopaedic implants by additive manufacturing using electron beam melting. *International Journal of Biomaterials*, 2012, 1–4. https://onlinelibrary.wiley.com/doi/10.1155/2012/245727

Pérez, M., Carou, D., Rubio, E. M., & Teti, R. (2020). Current advances in additive manufacturing. *Procedia CIRP*, 88, 439–444. www.sciencedirect.com/science/article/pii/S2212827120303978

Picard, M., Mohanty, A. K., & Misra, M. (2020). Recent advances in additive manufacturing of engineering thermoplastics: Challenges and opportunities. *RSC Advances*, 10(59), 36058–36089. https://pubs.rsc.org/en/content/articlelanding/2020/ra/d0ra04857g

Rengier, F., Mehndiratta, A., von Tengg-Kobligk, H. (2010). 3D printing based on imaging data: Review of medical applications. *IntJCARS*, 5, 335–341. https://pubmed.ncbi.nlm.nih.gov/20467825/

Revilla-León, M., Sadeghpour, M., & Özcan, M. (2020). A review of the applications of additive manufacturing technologies used to fabricate metals in implant dentistry. *Journal of Prosthodontics*, 29, 579–593. https://pubmed.ncbi.nlm.nih.gov/32548890/

Thompson, A., McNally, D., Maskery, I., & Leach, R. K. (2017). X-ray computed tomography and additive manufacturing in medicine: A review. *International Journal of Metrology and Quality Engineering*, 8, 17. www.metrology-journal.org/articles/ijmqe/full_html/2017/01/ijmqe170008/ijmqe170008.html

Trivedi, M., Jee, J., Silva, S., Blomgren, C., Pontinha, V. M., Dixon, D. L., ... & Gilmer, E. (2018). Additive manufacturing of pharmaceuticals for precision medicine applications: A review. *Additive Manufacturing,* 23, 319–328. www.sciencedirect.com/science/article/abs/pii/S221486041830335X

Vafadar, A., Guzzomi, F., Rassau, A., & Hayward, K. (2021). Advances in metal additive manufacturing: A review of common processes, industrial applications, and current challenges. *Applied Sciences*, 11(3), 1213. www.mdpi.com/2076-3417/11/3/1213

Ventola, C. L. (2014, October). Medical applications for 3D printing: Current and projected uses. P T. *Pharmacy and Therapeutics*, 39(10), 704–711. PMID: 25336867; PMCID: PMC4189697. https://pubmed.ncbi.nlm.nih.gov/25336867/

Yin, R. K. (2014). Case study research design and methods (5th ed.). *Canadian Journal of Program Evaluation*, 30(1), 108–110. https://doi.org/10.3138/cjpe.30.1.108

Zhou, X., Feng, Y., Zhang, J., Shi, Y., & Wang, L. (2020). Recent advances in additive manufacturing technology for bone tissue engineering scaffolds. *International Journal of Advanced Manufacturing Technology*, 108, 3591–3606. https://link.springer.com/article/10.1007/s00170-020-05444-1

Index

For Product Safety Concerns and Information please contact our EU
representative GPSR@taylorandfrancis.com
Taylor & Francis Verlag GmbH, Kaufingerstraße 24, 80331 München, Germany

www.ingramcontent.com/pod-product-compliance
Ingram Content Group UK Ltd.
Pitfield, Milton Keynes, MK11 3LW, UK
UKHW020159200926
475702UK00004B/535
9781041078531